LECTURE NOTES

ON

WAVES AND INSTABILITIES IN PLASMAS

LIU CHEN

PLASMA PHYSICS LABORATORY AND

DEPARTMENT OF ASTROPHYSICAL SCIENCES

PRINCETON UNIVERSITY

PRINCETON, NEW JERSEY

World Scientific Lecture Notes in Physics, Vol. 12

WAVES AND INSTABILITIES IN PLASMAS

by **Liu Chen**

World Scientific
Singapore • New Jersey • Hong Kong

Published by

World Scientific Publishing Co. Pte. Ltd.
P.O. Box 128, Farrer Road, Singapore 9128

U.S.A. office: World Scientific Publishing Co., Inc.
687 Hartwell Street, Teaneck NJ 07666, USA

Library of Congress Cataloging-in-Publication data is available.

WAVES AND INSTABILITIES IN PLASMAS

ISBN 9971-50-389-1
9971-50-390-5 pbk

Printed in Singapore by Kim Hup Lee Printing Co. Pte. Ltd.

TO SHINGSHAH AND PEIJIN

PREFACE

These notes have been given over the past several years in an one-semester course to second-year graduate students in the Plasma Physics Section of the Department of Astrophysical Sciences of Princeton University. The students generally have (or assumed to have) some basic knowledge of plasma dynamics in terms of single-particle and fluid descriptions. The topics covered in these notes, therefore, are selective and tend to emphasize more on kinetic-theory approaches to waves and instabilities in both uniform and nonuniform plasmas.

This work was in part supported by The Department of Energy and The National Science Foundation. I am thankful to Rosemary Fuchs, Terry Greenberg, Cecelia O'Brien, and Barbara Sarfaty for their patience and careful typing of the manuscript.

TABLE OF CONTENTS

CHAPTER I.

General Properties of Electromagnetic Waves in Dielectric Media

§I.1 Plasma as a dielectric medium

In this chapter, we shall discuss some general properties pertinent to linear electromagnetic waves in plasmas. In fact, these properties are sufficiently general and, hence, fundamental that they are applicable to other dielectric media.

Let us assume that the plasma or, equivalently, the dielectric medium to be time-stationary and spatially homogeneous. The perturbed (wave) electric field, $\delta \underset{\sim}{E}(\underset{\sim}{x}, t)$, can then be taken to assume the plane-wave form; i.e.,

$$\delta \underset{\sim}{E} (\underset{\sim}{x},t) = \delta \hat{\underset{\sim}{E}} \exp (i\omega t - i \underset{\sim}{k}\cdot\underset{\sim}{x}) \quad . \qquad (I.1.1)$$

Here, $\underset{\sim}{k}$ is the wave vector and ω is the frequency. Whether $(\omega,\underset{\sim}{k})$ is real or complex depends on the detailed nature of the problem. For example, $\omega=\omega_0$ is real if we are interested in wave properties driven by an external source oscillating at frequency ω_0. $\underset{\sim}{k}$, however, can become complex as in the case of evanescent waves. On the otherhand, if we are interested in an initial value problem (e.g., stability analysis), ω is then complex. This point can be realized by interpreting Eq. (I.1.1) as the Laplace-in-time and Fourier-in-space transform. Thus, we have, in this case, Im $\omega > 0$ to ensure causality (i.e., $\delta \underset{\sim}{E} \to 0$ as $t \to -\infty$) and $\underset{\sim}{k}$ is real.

In response to $\delta \underset{\sim}{E}$, charged particles will jiggle and acquire velocities, $\delta \underset{\sim}{V}$, and, hence, there will be perturbed current densities, $\delta \underset{\sim}{J} (\underset{\sim}{x},t)$. Since we are only interested in linear effects and the background plasma is taken to be stationary and homogeneous, it is quite general that

$$\delta\hat{\underset{\sim}{J}}\,(\omega,\underset{\sim}{k}) = \underset{\approx}{\sigma}\,(\omega,\underset{\sim}{k}) \cdot \delta\hat{\underset{\sim}{E}}\,(\omega,\underset{\sim}{k}) \quad . \tag{I.1.2}$$

Here, $\underset{\approx}{\sigma}$ is the conductivity tensor. For the present discussion, we shall simply assume $\underset{\approx}{\sigma}$ and Eq. (I.1.2) are given.

Substituting Eqs. (I.1.1) and (I.1.2) into the following Maxwell's equations

$$\underset{\sim}{\nabla} \times \underset{\sim}{E} = - \ (1/c)\partial\underset{\sim}{B}/\partial t \ , \tag{I.1.3}$$

$$\underset{\sim}{\nabla} \times \underset{\sim}{B} = (4\pi\underset{\sim}{j} + \partial\underset{\sim}{E}/\partial t)/c \ ; \tag{I.1.4}$$

and noting that $\underset{\sim}{E} = \underset{\sim}{E}_o + \delta\underset{\sim}{E}$, $\underset{\sim}{\nabla} \times \underset{\sim}{E}_o = 0$, $\partial\underset{\sim}{E}_o/\partial t = 0$ and etc., we find

$$\underset{\sim}{k} \times \delta\hat{\underset{\sim}{E}} = \omega\ \delta\hat{\underset{\sim}{B}}/c \tag{I.1.5}$$

$$\underset{\sim}{k} \times \delta\hat{\underset{\sim}{B}} = - \ (\omega/c)\ [\underset{\approx}{I} + i4\pi\ \underset{\approx}{\sigma}/\omega]\cdot\delta\hat{\underset{\sim}{E}}$$

$$\equiv - \ (\omega/c)\ \underset{\approx}{D} \cdot \delta\hat{\underset{\sim}{E}} \quad . \tag{I.1.6}$$

Here, $\underset{\approx}{I}$ is the unit tensor and $\underset{\approx}{D} \equiv \underset{\approx}{I} + i4\pi\ \underset{\approx}{\sigma}/\omega$ is the equivalent dielectric tensor. Combining Eqs. (I.1.5) and (I.1.6), we obtain

$$\left[\underset{\sim}{k}\,\underset{\sim}{k} - k^2\,\underset{\approx}{I} + \frac{\omega^2}{c^2}\,\underset{\approx}{D}\right] \cdot \delta\hat{\underset{\sim}{E}} \equiv \underset{\approx}{\varepsilon} \cdot \delta\hat{\underset{\sim}{E}} = 0 \quad . \tag{I.1.7}$$

In deriving Eq. (I.1.7) , we have used the vector identity

$$\underset{\sim}{\nabla} \times (\underset{\sim}{\nabla} \times \underset{\sim}{A}) = \underset{\sim}{\nabla} (\underset{\sim}{\nabla} \cdot \underset{\sim}{A}) - \nabla^2 \underset{\sim}{A} .$$

In order that Eq. (I.1.7) has non-trival solution of $\delta\hat{\underset{\sim}{E}}$, the determinant of $\underset{\approx}{\varepsilon}$ must vanish; i.e.,

$$\varepsilon(\omega,\underset{\sim}{k}) \equiv \| \underset{\sim}{\varepsilon}(\omega,\underset{\sim}{k}) \| = 0 \quad . \tag{I.1.8}$$

Equation (I.1.8) is the so-called dispersion relation. That is, from Eq. (I.1.8) we can obtain a relation between ω and $\underset{\sim}{k}$, $\omega = \omega(\underset{\sim}{k})$, and, thus, plot the ω vs. $\underset{\sim}{k}$ dispersion curves.

In general, given a $\underset{\sim}{k}$, the dispersion relation can contain multiple (or, sometimes, infinite) number of ω roots. Indexing these ω branches by j, we then have

$$\omega = \omega_j(\underset{\sim}{k}) \text{ for } j = 1,2..... \quad . \tag{I.1.9}$$

In plasma physics literatures, the various ω branches are sometimes called normal modes. Substituting Eq. (I.1.9) into Eq. (I.1.7), we can then determine the eigenvector $\delta\hat{\underset{\sim}{E}}_j$ corresponding to the eigenvalue ω_j. From $\delta\hat{\underset{\sim}{E}}_j$ the polarization of the perturbed electric (or magnetic) field for the jth branch (normal mode) with respect to $\underset{\sim}{k}$ or the equilibrium magnetic field $\underset{\sim}{B}_o$ is then known.

<u>Homework #1</u>

(1) Illustrate with an example the above mentioned concepts: conductivity tensor, dispersion relation, ω branches (or modes), eigenvector $\delta\underset{\sim}{E}$ and wave polarization.

(2) Consider the following Inverse Laplace transform $\phi(t) = \int_c (d\omega/2\pi) \times e^{-i\omega t}/D(\omega)$. How will you choose the integration contour c to ensure causality for (i) D (ω) has no zeroes in the upper-half (Im ω>0) plane? (ii) D (ω) has finite number of zeroes in the Im $\omega > 0$ plane?

(3) (i) what is the condition that $\delta\phi(x,t)$ is Fourier transformable in x; i.e., $\delta\phi(k,t) = \int dx\ \delta\phi(x,t)e^{-ikx}$ exists? (ii) Suppose that $\delta\phi$ at $t = 0$ is sufficiently smooth and of finite spatial extent; i.e., $\delta\phi(x,t=0) = 0$ for $|x| \geq d$. Is $\delta\phi(x,t)$ Fourier transformable for any finite t and why?

§I.2 Two time-scale analysis

In plasma, due to the presence of weak dissipation or source, waves tend to either grow or decay slowly with time. In many cases of interest, the dissipation or source is sufficiently weak such that the characteristic growth or decay rate is much smaller than the typical oscillation frequency ω_o. The wave process is then said to possess two time scales and we may assume δE to be of the following form

$$\delta E\ (t) = \delta\hat{E}\ (\epsilon t)\ \exp\ (-i\omega_o t). \qquad (I.2.1)$$

Here, for simplicity of discussions, we have neglected spatial dependence and taken δE to be a scalar and $\epsilon \ll 1$ is a small parameter. Thus,

$$|\partial\delta\hat{E}/\partial t|/|\omega_o\delta\hat{E}| \sim o(\epsilon) \ll 1. \qquad (I.2.2)$$

Using Eqs. (I.2.1) and (I.2.2), we shall explore some implications of the two time scales. First, we show that given the relation

$$\delta A(\omega) = B(\omega)\delta E(\omega) \tag{I.2.3}$$

with ω being, strictly speaking, the Laplace-transform variable, we have

$$\delta\hat{A}(\varepsilon t) = B\,(\omega_0 + i\,\frac{\partial}{\partial t})\;\delta\hat{E}(\varepsilon t). \tag{I.2.4}$$

Proof of Eq. (I.2.4): From Eq. (I.2.3), we have

$$\delta A(t) = \int \delta A(\omega)\, e^{-\,i\omega t}\, d\omega = \int d\omega\; B\omega \int \frac{dt'}{2\pi}\;\delta E(t')e^{i\omega(t'-t)} \tag{I.2.5}$$

Taylor expanding $B(\omega)$ about $\omega=\omega_0$ and noting Eq. (I.2.1), Eq. (I.2.5) becomes

$$\delta A(t) = e^{-i\omega_0 t} \int \frac{dt'}{2\pi}\;\delta\hat{E}(\varepsilon t') \int d\omega\;[B(\omega_0) + \frac{\partial B}{\partial \omega_0}\;(\omega-\omega_0)\ldots]\; e^{i(\omega-\omega_0)(t'-t)}$$

$$= e^{-i\omega_0 t} \int \frac{dt'}{2\pi}\;\delta\hat{E}(\varepsilon t')[B(\omega_0) + \frac{\partial B}{\partial \omega_0}\,\frac{i\partial}{\partial t} +\ldots] \int d\omega e^{i(\omega-\omega_0)(t'-t)}$$

$$= e^{-i\omega_0 t}[B(\omega_0) + \frac{\partial B}{\partial \omega_0}\; i\,\frac{\partial}{\partial t} + \ldots] \int dt'\;\delta\hat{E}(\varepsilon t')\delta(t-t) \quad ;$$

that is,

$$\delta\hat{A}\;(\varepsilon t) = [B(\omega_0) + \frac{\partial B}{\partial \omega_0}\; i\,\frac{\partial}{\partial t} + \ldots]\;\delta\hat{E}(\varepsilon t). \tag{I.2.4}$$

Equation (I.2.4) can be derived alternatively by noting that

$$L_p^{-1}[-\,i\omega\;\delta E(\omega)] = \frac{\partial}{\partial t}\;\delta E(t) = -\;ie^{-i\omega_0 t}(\omega_0 + i\,\frac{\partial}{\partial t})\delta\hat{E}\;(\varepsilon t). \tag{I.2.6}$$

Thus regarding ω as an operator, Eq. (I.2.3) inversely transforms to

$$\delta \underset{\sim}{E}(t) = [\, \delta \hat{\underset{\sim}{E}}(\varepsilon t) e^{-i\omega_o t} + c.c.]/2 \tag{I.2.13}$$

and similarly for $\delta \underset{\sim}{B}(t)$, we find, using Eq. (I.2.4)

$$4\pi\ \delta \underset{\sim}{J} + \frac{\partial}{\partial t}\, \delta \underset{\sim}{E} = (1/2) e^{-i\omega_o t} [-i(\omega_o + i\frac{\partial}{\partial t}) \underset{\approx}{D}(\omega_o + i\frac{\partial}{\partial t}) \cdot \delta \hat{\underset{\sim}{E}}]$$

$$+ (1/2) e^{i\omega_o t} [(-\, i)(-\omega_o + i\frac{\partial}{\partial t}) \underset{\approx}{D}(-\omega_o + i\frac{\partial}{\partial t}) \cdot \delta \hat{\underset{\sim}{E}}^*]$$

$$\approx (1/2)\{e^{-i\omega_o t} [-\, i\omega_o \underset{\approx}{D}(\omega_o) + \frac{\partial}{\partial \omega_o} [\omega_o \underset{\approx}{D}(\omega_o)] \frac{\partial}{\partial t}] \cdot \delta \hat{\underset{\sim}{E}}$$

$$+ e^{i\omega_o t} [i\omega_o \underset{\approx}{D}(-\omega_o) + \frac{\partial}{\partial \omega_o} [\omega_o \underset{\approx}{D}(-\omega_o)] \frac{\partial}{\delta t}] \cdot \delta \hat{\underset{\sim}{E}}^* \}. \tag{I.2.14}$$

Substituting Eq. (I.2.14) into Eq. (I.2.11) and performing the phase averaging, we have, term by term,

$$\overline{\delta \underset{\sim}{E} \cdot (4\pi \delta \underset{\sim}{J} + \frac{\partial}{\partial t}\, \delta \underset{\sim}{E})} = (1/4)\ \{-\, i\omega_o \delta \hat{\underset{\sim}{E}}^* \cdot \underset{\approx}{D}(\omega_o) \cdot \delta \hat{\underset{\sim}{E}} + i\omega_o\ \delta \hat{\underset{\sim}{E}} \cdot \underset{\approx}{D}(-\omega_o) \cdot \delta \hat{\underset{\sim}{E}}^* +$$

$$\delta \hat{\underset{\sim}{E}}^* \cdot \frac{\partial}{\partial \omega_o} [\omega_o \underset{\approx}{D}(\omega_o)] \frac{\partial}{\partial t}\, \delta \hat{\underset{\sim}{E}} + \delta \hat{\underset{\sim}{E}} \cdot \frac{\partial}{\partial \omega_o} [\omega_o \underset{\approx}{D}(-\omega_o)] \cdot \frac{\partial}{\partial t}\, \delta \hat{\underset{\sim}{E}}^* \}. \tag{I.2.15}$$

Defining

$$\underset{\approx}{D} = \underset{\approx}{D}_h + i\, \underset{\approx}{D}_a,$$

where h and a denote, respectively, Hermitian and anti-Hermitan components; i.e.,

$$\underset{\approx}{D}_h(\omega_o) = (1/2)\{\underset{\approx}{D} + (\underset{\approx}{D}^T)^*\}\ , \tag{I.2.17}$$

and

$$\underset{\approx}{D}_a(\omega_o) = (1/2i)\{\underset{\approx}{D} - (\underset{\approx}{D}^T)^*\}. \qquad \text{(I.2.18)}$$

Here $(A_{ij})^{T*} = A^*_{ji}$; i.e., the complex conjugate of the transpose of $\underset{\approx}{A}$. We note, futhermore, that from realizability condiion

$$\underset{\approx}{D}(-\omega_o) = \underset{\approx}{D}^*(\omega_o). \qquad \text{(I.2.19)}$$

The proof of Eq. (I.2.19) is straightforward by noting that both $\delta\underset{\sim}{E}$ and $\partial\delta\underset{\sim}{E}/\partial t + 4\pi\delta\underset{\sim}{J}$ are real physical quantities. Equation (I.2.15) then reduces to, assuming $|\underset{\approx}{D}_a| << |\underset{\approx}{D}_h|$,

$$-i\omega_o\delta\hat{E}^*_j D_{ji}\delta\hat{E}_i + i\omega_o\,\delta\hat{E}_i D^*_{ij}\,\delta E_j^* + \delta\hat{E}^*_j \frac{\partial}{\partial\omega_o}[\omega_o D_{ji}]\frac{\partial}{\partial t}\delta\hat{E}_i$$

$$+\,\delta\hat{E}_i\frac{\partial}{\partial\omega_o}[\omega_o D^*_{ij}]\frac{\partial}{\partial t}\delta\hat{E}^*_j = 2\omega_o\delta\hat{E}^*_j(D_{ji})_a\,\delta\hat{E}_i + \frac{\partial}{\partial t}[\delta\hat{E}^*_j\frac{\partial}{\partial\omega_o}(\omega_o D_{ji})_h\delta\hat{E}_i]. \qquad \text{(I.2.20)}$$

All together, phase averaging Eq. (I.2.11) leads to

$$\frac{\partial}{\partial t}\delta\hat{W} + \frac{1}{8\pi}\omega_o\,\delta\hat{\underset{\sim}{E}}^*\cdot\underset{\approx}{D}_a\cdot\delta\hat{\underset{\sim}{E}} = 0, \qquad \text{(I.2.21)}$$

where

$$\delta\hat{W} = \frac{1}{16\pi}[|\delta\hat{\underset{\sim}{B}}|^2 + \delta\hat{\underset{\sim}{E}}^*\cdot\frac{\partial}{\partial\omega_o}(\omega_o\,\underset{\approx}{D}_h)\cdot\delta\hat{\underset{\sim}{E}}] \qquad \text{(I.2.22)}$$

is the wave energy. We note that $\delta\hat{W}$ consists of three components: the magnetic field energy $|\delta\hat{\underset{\sim}{B}}|^2$, the electric field energy and coherent particle kinetic (mechanical) energy. The latter two contributions are combined into the second term in Eq. (I.2.22). This is clear by noting Eq. (I.1.6) (i.e., $\underset{\approx}{D} = \underset{\approx}{I} + i4\pi\underset{\approx}{\sigma}/\omega$). Thus,

$$\delta\hat{\underset{\sim}{E}}^* \cdot \frac{\partial}{\partial\omega_o}(\omega_o \underset{\approx}{D}_h)\cdot\delta\hat{\underset{\sim}{E}}^* = |\delta\hat{\underset{\sim}{E}}|^2 + \delta\hat{\underset{\sim}{E}}^* \cdot \frac{\partial}{\partial\omega_o}(i4\pi\underset{\approx}{\sigma})_h \cdot \delta\hat{\underset{\sim}{E}} \ . \tag{I.2.23}$$

Since $\delta\underset{\sim}{J} = \underset{\approx}{\sigma} \cdot \delta\underset{\sim}{E}$, which is due to particle coherent dynamics, the meaning of particle kinetic (mechanical) energy is transparent.

Equation (I.2.21) clearly shows that in plasmas wave energy may either decrease or increase depending on the anti-Hermitian part of $\underset{\approx}{D}$. As a simple example, we take $\underset{\sim}{k}=0$ and $\underset{\approx}{D} = D_s \underset{\approx}{I}$. We then have, noting $\delta\underset{\sim}{B}=0$,

$$\frac{1}{16\pi}\frac{\partial}{\partial t}\left[\frac{\partial}{\partial\omega_o}(\omega_o D_{sh})|\delta\hat{\underset{\sim}{E}}|^2\right] + \frac{1}{8\pi}\omega_o D_{sa}|\delta\hat{E}|^2 = 0. \tag{I.2.24}$$

Equation (I.2.24) has the solution

$$|\delta\hat{\underset{\sim}{E}}|^2 = |\delta\hat{\underset{\sim}{E}}_o|^2 \exp(2\gamma t) \tag{I.2.25}$$

with

$$\gamma/\omega_o = - D_{sa}/[\partial(\omega_o D_{sh})/\partial\omega_o]. \tag{I.2.26}$$

Noting also that in this case $D_{sh} = \mathrm{Re}(D_s) \equiv D_{sr}$ and $D_{sa} = D_{si}$. Thus, for waves with positive wave energy $\partial(\omega_o D_{sr})/\partial\omega_o > 0$, we would have wave growth (decay) if the dissipation, i.e., $\omega_o D_{si}$, is negative (positive). The converse is true for negative-energy waves. Finally, Eqs. (I.2.21) and (I.2.22) indicate that the consistency of the two-time-scale analysis requires the previously stated assumption

$$|\underset{\approx}{D}_a| \;/\; |\underset{\approx}{D}_h| \sim O(\varepsilon) << 1. \tag{I.2.27}$$

§I.3 Analysis with two time and space scales

In plasmas, spatial dependence plays a crucial role. In particular, it gives rise to the wave dispersion; i.e., ω depends on the wave vector $\underset{\sim}{k}$. As in the case of temporal variations, wave spatial variations also often exhibits two scales, one rapid and one slow. Such two-space-scale phenomena can occur due to various reasons; e.g., dissiation or source, propagation of wave packet, weak nonuniformity, etc.

Thus, in general, we would allow fast and slow variations in both time and space. $\delta\underset{\sim}{E}(\underset{\sim}{x},t)$ can then be expressed in the following form

$$\delta\underset{\sim}{E}(t,\underset{\sim}{x}) = \delta\underset{\sim}{E}(\varepsilon t,\varepsilon\underset{\sim}{x}) \exp(i\underset{\sim}{k}_o \cdot \underset{\sim}{x} - i\omega_o t). \tag{I.3.1}$$

We can then carry out analyses similar to those done in Sec. §I.2. For example, given the following relation in $(\omega,\underset{\sim}{k})$ space

$$\delta\underset{\sim}{A}(\omega,\underset{\sim}{k}) = \underset{\approx}{B}\,(\omega,\underset{\sim}{k})\cdot\delta\underset{\sim}{E}(\omega,\underset{\sim}{k}) \quad ; \tag{I.3.2}$$

where $(\omega,\underset{\sim}{k})$ are the Laplace and Fourier transform variables, we can find, in the $(t,\underset{\sim}{x})$ space, that

$$\delta\hat{\underset{\sim}{A}}(\varepsilon t,\varepsilon\underset{\sim}{x}) = [\underset{\approx}{B}(\omega_o + i\,\frac{\partial}{\partial t},\, \underset{\sim}{k}_o - i\,\frac{\partial}{\partial \underset{\sim}{x}})]\cdot\delta\hat{\underset{\sim}{E}}\,(\varepsilon t,\, \varepsilon\underset{\sim}{x}). \tag{I.3.3}$$

Expanding Eq. (I.3.3) to first order, we obtain

$$\delta\hat{A}_i(\epsilon t,\epsilon \underset{\sim}{x}) \cong [B_{ij}(\omega_o, k_{o\ell}) + \frac{\partial B_{ij}(\omega_o, k_{o\ell})}{\partial\omega_o}(i\frac{\partial}{\partial t})$$

$$+ \frac{\partial B_{ij}(\omega_o,k_{o\ell})}{\partial k_{om}}(-i\frac{\partial}{\partial x_m})]\delta\hat{E}_j(\epsilon t,\epsilon \underset{\sim}{x}). \qquad (I.3.4)$$

Here, we have used indicial tensor representation to clarify the operations. We can of course carry out a similar phase averaging both in time and space. This is straightforward and we shall not discuss it.

In stead, we shall substitute Eqs. (I.3.1) and (I.3.4) into Maxwell's equations, Eqs. (I.1.5) and (I.1.6), and examine the results order by order. For this purpose, we note that

$$\underset{\approx}{D} = \underset{\approx}{D}_h + i\,\underset{\approx}{D}_a \qquad (I.3.5)$$

and assume, more specifically,

$$|\delta\hat{\underset{\sim}{E}}^* \cdot [\partial(\omega_o \underset{\approx}{D}_a/\partial\omega_o] \cdot \delta\hat{\underset{\sim}{E}}| / |\delta\hat{\underset{\sim}{E}}^* \cdot [\partial(\omega_o \underset{\approx}{D}_h)/\partial\omega_o] \cdot \delta\hat{\underset{\sim}{E}}| \sim O(\epsilon) << 1. \qquad (I.2.27)'$$

In the zeroth order, O(1), we obtain the expected result

$$\underset{\approx}{\epsilon}_h(\omega_o,\underset{\sim}{k}_o) \cdot \delta\hat{\underset{\sim}{E}} = [\underset{\sim}{k}_o\underset{\sim}{k}_o - k_o^2\,\underset{\approx}{I} + (\omega_o/c)^2\underset{\approx}{D}_h] \cdot \delta\hat{\underset{\sim}{E}} = 0. \qquad (I.3.6)$$

Equation (I.3.6) gives the zeroth-order dispersion relation

$$\epsilon_h(\omega_o,\underset{\sim}{k}_o) \equiv \| \underset{\approx}{\epsilon}_h(\omega_o, \underset{\sim}{k}_o) \| = 0. \qquad (I.3.7)$$

An alternative expression for the dispersion relation can be obtained by dotting $\delta\hat{\underset{\sim}{E}}^*$ into Eq. (I.3.6) to obtain

$$|\underset{\sim}{k}_o \times \delta\hat{\underset{\sim}{E}}|^2 = (\omega_o/c)^2\ \delta\hat{\underset{\sim}{E}}^* \cdot \underset{\approx}{D}_h \cdot \delta\hat{\underset{\sim}{E}}. \tag{I.3.8}$$

In the next order, $O(\varepsilon)$, we have

$$c\,\frac{\partial}{\partial \underset{\sim}{x}} \times \delta\hat{\underset{\sim}{E}} = -\frac{\partial}{\partial t}\,\delta\hat{\underset{\sim}{B}} \tag{I.3.9}$$

$$c\,\frac{\partial}{\partial \underset{\sim}{x}} \times \delta\hat{\underset{\sim}{B}} = \omega_o\,\underset{\approx}{D}_a \cdot \delta\hat{\underset{\sim}{E}} + \frac{\partial}{\partial\omega_o}(\omega_o \underset{\approx}{D}_h)\,\frac{\partial}{\partial t}\,\delta\hat{\underset{\sim}{E}} - \omega_o\,\frac{\partial}{\partial \underset{\sim}{k}_o}\underset{\approx}{D}_h \cdot \frac{\partial}{\partial \underset{\sim}{x}}\,\delta\hat{\underset{\sim}{E}}\ . \tag{I. 3.10}$$

Now we perform the following operations

$$\delta\hat{\underset{\sim}{E}}^* \cdot \text{Eq. (I.3.10)} + \delta\hat{\underset{\sim}{E}} \cdot \text{Eq. (I.3.10)}^* - \delta\hat{\underset{\sim}{B}}^* \cdot \text{Eq. (I.3.9)} - \delta\hat{\underset{\sim}{B}} \cdot \text{Eq.}^* \text{(I.3.9)}. \tag{I.3.11}$$

We obtain, noting that $\underset{\approx}{D}(-\omega_o, -\underset{\sim}{k}_o) = \underset{\approx}{D}^*(\omega_o, \underset{\sim}{k}_o)$,

$$\frac{\partial}{\partial t}\left[|\delta\hat{\underset{\sim}{B}}|^2 + \delta E^*\,\frac{\partial}{\partial\omega_o}(\underset{\approx}{D}_h\omega_o)\cdot\delta\hat{\underset{\sim}{E}}\right] + 2\omega_o\delta\hat{\underset{\sim}{E}}^* \cdot \underset{\approx}{D}_a \cdot \delta\hat{\underset{\sim}{E}} - \omega_o\,\frac{\partial}{\partial \underset{\sim}{x}} \cdot \frac{\partial}{\partial \underset{\sim}{k}_o}(\delta\hat{\underset{\sim}{E}}^* \cdot \underset{\approx}{D}_h \cdot \delta\hat{\underset{\sim}{E}})$$

$$= -c\,\frac{\partial}{\partial \underset{\sim}{x}} \cdot [\delta\hat{\underset{\sim}{E}}^* \times \delta\hat{\underset{\sim}{B}} + \delta\hat{\underset{\sim}{E}} \times \delta\hat{\underset{\sim}{B}}^*]. \tag{I.3.12}$$

From Eq. (I.3.8), we note that

$$|\delta\hat{\underset{\sim}{B}}|^2 = \left(\frac{c}{\omega_o}\right)^2 |\underset{\sim}{k}_o \times \delta\hat{\underset{\sim}{E}}|^2 = \delta\hat{\underset{\sim}{E}}^* \cdot \underset{\approx}{D}_h \cdot \delta\hat{\underset{\sim}{E}}. \tag{I.3.13}$$

Equation (I.3.12) divided by 16π can then be written as

$$\frac{\partial}{\partial t}\,\delta W - \frac{\omega_o}{16\pi}\frac{\partial}{\partial \underset{\sim}{x}}\cdot\frac{\partial}{\partial \underset{\sim}{k}_o}(\delta\hat{\underset{\sim}{E}}^*\cdot\underset{\approx}{D}_h\cdot\delta\hat{\underset{\sim}{E}}) + \frac{c}{8\pi}\frac{\partial}{\partial \underset{\sim}{x}}\cdot R_e(\delta\hat{\underset{\sim}{E}}^*\times\delta\hat{\underset{\sim}{B}}) +$$

$$\frac{\omega_o}{8\pi}\,\delta\hat{\underset{\sim}{E}}^*\cdot\underset{\approx}{D}_a\cdot\delta\hat{\underset{\sim}{E}} = 0, \qquad (I.3.14)$$

where

$$\delta W \equiv \frac{1}{16\pi}\,\delta\hat{\underset{\sim}{E}}^*\cdot\frac{\partial}{\partial\omega_o}(\omega_o^{\,2}\,\underset{\approx}{D}_h)\cdot\delta\hat{\underset{\sim}{E}}/\omega_o\,. \qquad (I.3.15)$$

Now, let us define

$$\underset{\sim}{V}_{gr} \equiv -\,[\,\omega_o\frac{\partial}{\partial\underset{\sim}{k}_o}(\delta\hat{\underset{\sim}{E}}^*\cdot\underset{\approx}{D}_h\cdot\delta\hat{\underset{\sim}{E}}) - 2c\,\mathrm{Re}(\delta\hat{\underset{\sim}{E}}^*\times\delta\hat{\underset{\sim}{B}})]/16\pi\delta W$$

$$= -\,\{2c^2[\underset{\sim}{k}_o|\delta\hat{\underset{\sim}{E}}|^2 - \mathrm{Re}\,(k_o\cdot\delta\hat{\underset{\sim}{E}}^*)\delta\hat{\underset{\sim}{E}}] - \omega_o^{\,2}\frac{\partial}{\partial\underset{\sim}{k}_o}(\delta\hat{\underset{\sim}{E}}^*\cdot\underset{\approx}{D}_h\cdot\delta\hat{\underset{\sim}{E}})\}/\{$$

$$\frac{\partial}{\partial\omega_o}(\omega_o^{\,2}\delta\hat{\underset{\sim}{E}}^*\cdot\underset{\approx}{D}_h\cdot\delta\hat{\underset{\sim}{E}})\}. \qquad (I.3.16)$$

Equation (I.3.14) then becomes the familiar form

$$\frac{\partial}{\partial t}\,\delta W + \frac{\partial}{\partial\underset{\sim}{x}}\cdot(\underset{\sim}{V}_{gr}\,\delta W) = 2\omega_i\,\delta W, \qquad (I.3.17)$$

where

$$\omega_i \equiv -\,[\frac{\omega_o}{16\pi}\,\delta\hat{\underset{\sim}{E}}^*\cdot\underset{\approx}{D}_a\cdot\delta\hat{\underset{\sim}{E}}]/\delta W. \qquad (I.3.18)$$

$\underset{\sim}{V}_{gr}$ as defined by Eq. (I.3.16) can be interpreted as

$$\underset{\sim}{V}_{gr} = \text{total wave energy flux/wave energy}. \qquad (I.3.19)$$

Meanwhile, since

$$\underset{\sim}{V}_{gr} = \partial\omega_o/\partial\underset{\sim}{k}_o, \tag{I.3.20}$$

it shows that the wave energy is carried by the wave packet or, sometimes called, quasi particle which propagates at the velocity $\underset{\sim}{V}_{gr}$. If $\omega_i = 0$; i.e. the plasma is Hermitian, then the wave energy remains constant along the propagation path of the wave packet. With $\omega_i \neq 0$, the wave energy can either increase or decrease along the path.

Homework #2

(1) Using the physical realizability condition, prove that

$$\underset{\approx}{D}\,(-\omega) = \underset{\approx}{D}^*\,(\omega^*) \quad . \tag{H.2.1}$$

(2) The cold plasma Langmuir oscillation with collisional dissaption is governed by the following equations

$$\frac{\partial}{\partial t}\,\delta\underset{\sim}{V} = -\frac{e}{m}\,\delta\underset{\sim}{E} - \nu\delta\underset{\sim}{V} \tag{H.2.2}$$

$$\frac{\partial}{\partial t}\,\delta\underset{\sim}{E} = 4\pi N_o e\,\delta\underset{\sim}{V}\ ; \tag{H.2.3}$$

or, combining Eqs. (H.2.2) and (H.2.3)

$$\frac{\partial^2}{\partial t^2}\,\delta\underset{\sim}{V} = -\,\omega_{pe}{}^2\,\delta\underset{\sim}{V} - \nu\,\frac{\partial}{\partial t}\,\delta\underset{\sim}{V} \quad . \tag{H.2.4}$$

Here, $\omega_{pe}^2 = 4\pi N_o e^2/m_e$ and we shall assume $|\nu| << |\omega_{pe}|$.

(2.a) Solving Eq. (H.2.4) for $\delta \underset{\sim}{V}$ using the two-time scale approach.

(2.b) Derive $\underset{\approx}{D}$ and indicate the Hermitian and anti-Hermitian parts.

(2.c) Derive an expression for the wave energy δW.

(2.d) Calculate the damping or growth rate using δW and $\underset{\approx}{D}_a$.

(3) Using the disperison relation given by Eq. (I.3.8), prove that the expression of $\underset{\sim}{V}_{gr}$ given by Eq. (I.3.16) agrees with the more familiar one, $\underset{\sim}{V}_{gr} = \partial\omega_o/\partial\underset{\sim}{k}_o$.

§I.4 Instabilities

Plasmas in both laboratories and space are often far from thermal equilibrium. For example, in mirror machines as well as earth's magnetosphere, the velocity distribution is non-Maxwellian due to the existence of loss cones. Furthermore, in any confined plasmas either inertially or magnetically, there is also nonuniformities in macroscopic thermodynamic quantities such as density, temperature, pressure, etc. In a way, these deviations from thermodynamic equilibrium may be regarded as free energies stored in the plasma. Conventionally, they may be catagorized in the following three types:

(i) Velocity-space free energy: velocity-space anisotropy, beams. Examples are loss-cone and two-stream instabilities.

(ii) Magnetic free energy: current and magnetic field inhomogeneities. Examples are kink and tearing instabilities.

(iii) Expansion free energy: density, temperature, pressure inhomogeneities.

Examples are drift, ballooning/interchange and trapped-particle instabilities.

Instabilities are collective processes which may be triggered (if certain threshold conditions are satisfied) to release the above-mentioned free energies. That is, via the instabilities, these free energies can be converted into either turbulent plasma motions and/or electromagnetic radiations. In this respect, instabilities may be viewed as anomalous (compared to collisional) processes whereby plasmas can relax toward thermal equilibrium and, therefore, instabilities play crucial roles in our understanding of important subjects in both laboratory and space plasmas; such as disruption in tokamaks, anomalous transport processes, collisionless shocks, solar radio bursts, etc.

In an infinite, uniform plasma, we say an instability exists if and only if for some <u>real</u> $\underset{\sim}{k}$, the dispersion relation

$$\varepsilon(\omega,\underset{\sim}{k}) \equiv |\underset{\sim}{\varepsilon}(\omega,\underset{\sim}{k})| = 0 \tag{I.4.1}$$

admits a solution with $\mathrm{Im}\,\omega \equiv \gamma > 0$. Here, we note our convention is

$$\delta\underset{\sim}{E}(\underset{\sim}{x},t) = \delta\hat{\underset{\sim}{E}} \exp(i\underset{\sim}{k} \cdot \underset{\sim}{x} - i\omega t)$$

with $(\omega,\underset{\sim}{k})$ properly understood to be the Laplace-in-t and Fourier-in-$\underset{\sim}{x}$ transform pair. That is, if a linear instability exists, a wave, which is <u>periodic</u> in space with a wave vector $\underset{\sim}{k}$, will exponentiate in time with the growth rate given by γ.

Based on the instability excitation mechanisms, we may divide instabilities into two types: (i) dissipative-type instabilities and (ii) reactive-type instabilities.

Dissipative-type instabilities are excited via (negative or positive) dissipative processes or, more generally speaking, the anti-Hermitian component of the dielectric tensor; as we have discussed in Sec. §I.2 and Sec. §I.3. Thus, we assume

$$\varepsilon(\omega,\underset{\sim}{k}) = \varepsilon_h(\omega,\underset{\sim}{k}) + i\varepsilon_a(\omega,\underset{\sim}{k}) = 0, \qquad (I.4.2)$$

and

$$|\varepsilon_a|/|\varepsilon_h| \sim O(\varepsilon) << 1. \qquad (I.4.3)$$

Furthermore, we let

$$\omega = \omega_r + i\gamma \text{ and } |\gamma/\omega_r| \sim O(\varepsilon) << 1.$$

Equation (I.4.2) then becomes, order by order, O(1):

$$\varepsilon_h(\omega_r,\underset{\sim}{k}) = 0 \qquad (I.4.4)$$

or

$$\omega_r = \omega_{rj}(\underset{\sim}{k}) \qquad \text{for } j = 1, \ldots , N , \qquad (I.4.5)$$

where N is the number of ω branches (i.e., normal modes). In the next order, we have, from Taylor expansion of ε_h about $\omega = \omega_{rj}$,

$$\gamma_j = - \varepsilon_a(\omega_{rj},\underset{\sim}{k})/[\frac{\partial}{\partial\omega_{rj}} \varepsilon_h(\omega_{rj},\underset{\sim}{k})] . \qquad (I.4.6)$$

Thus, for plasmas with negative dissipations ($\varepsilon_a < 0$), $\gamma_j > 0$ if $\partial\varepsilon_h/\partial\omega_{rj} > 0$; i.e., if the ($\omega_{rj}, \underset{\sim}{k}$) wave is a positive-energy wave. For a negative-energy wave ($\partial\varepsilon_h/\partial\omega_{rj} < 0$), however, we have $\gamma_j < 0$; the wave is stable or damped. Converse conclusions can be made for positive dissipations. In Homework #3, we shall have an example of a negative-energy wave driven unstable via positive dissipations.

As to reactive-type instabilities, the most well-known example is the beam-plasma instability. First, however, we shall illustrate with the following model dispersion relation

$$\varepsilon_h(\omega,k) = 1 - \frac{\omega_p}{\omega} + \frac{\omega_b}{(\omega - kv_o)} = 0. \tag{I.4.7}$$

Equation (I.4.7) is a quadratic equation in ω and, hence, can be readily solved. For simplicity, we shall assume $|\omega_b| \ll |\omega_p|$. Thus,

$$\omega_{1,2} \simeq \frac{(kv_o+\omega_p-\omega_b) \pm [(kv_o-\omega_p)^2 - 2\omega_b(kv_o+\omega_p)]^{1/2}}{2} \quad . \tag{I.4.8}$$

Thus, there exists an instability if

$$|kv_o - \omega_p| \lesssim 2(\omega_p\omega_b)^{1/2}, \tag{I.4.9}$$

and the maximum growth rate $\gamma_{max} \simeq (\omega_p\omega_b)^{1/2}$ occurs at $kv_o = \omega_p$. These results are sketched in terms of the ω-k dispersion curves shown in Fig. (I.4.1).

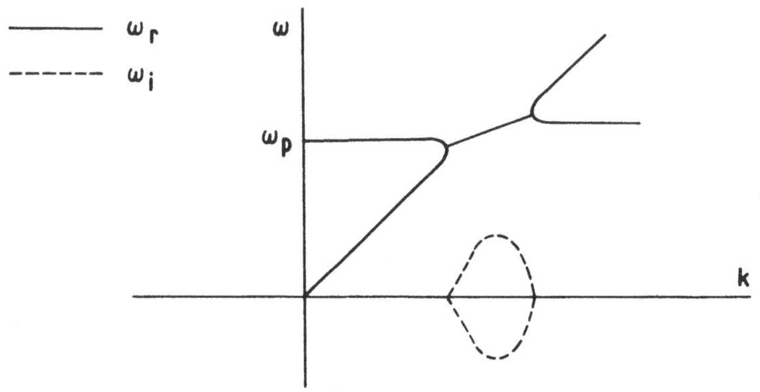

Fig. (I.4.1) Sketch of ω vs. k curves for the model dispersion relation of Eq. (I.4.7).

Since instability occurs at $\omega_{pe} \simeq kv_o$, this model reactive-type instability can be understood in terms of coupling between the $\omega_1 \simeq \omega_{pe}$ and $\omega_2 \simeq kv_o$ modes. Now, away from coupling, the corresponding dielectric constant for ω_1 and ω_2 modes are, respectively,

$$\varepsilon_{h1} = 1 - \omega_p/\omega_1 , \tag{I.4.10}$$

and

$$\varepsilon_{h2} = 1 + \omega_b/(\omega_2 - kv_o) . \tag{I.4.11}$$

since

$$\partial(\omega_1\varepsilon_{h1})/\partial\omega_1 = 1 > 0 , \tag{I.4.12}$$

the ω_1 mode is a positive-energy wave. On the other hand,

$$\partial(\omega_2\epsilon_{h2})/\partial\omega_2 \simeq 1 - \frac{kv_o\omega_b}{(\omega_2-kv_o)^2} \simeq -\frac{kv_o}{\omega_b} < 0 , \qquad (I.4.13)$$

i.e., the ω_2 mode is a negative-energy wave. Thus, the above model reactive-type instability can be explained in terms of coupling between a negative-energy and a positive-energy wave. It is easy to see that if both modes are positive-energy waves, then the system is stable. For example, the following model dispersion relation

$$\epsilon_h(\omega,k) = 1 - \frac{\omega_p}{\omega} - \frac{\omega_b}{(\omega-kv_o)} = 0 \qquad (I.4.14)$$

has no unstable solutions. The ω-k dispersion curve is given in Fig. (I.4.2).

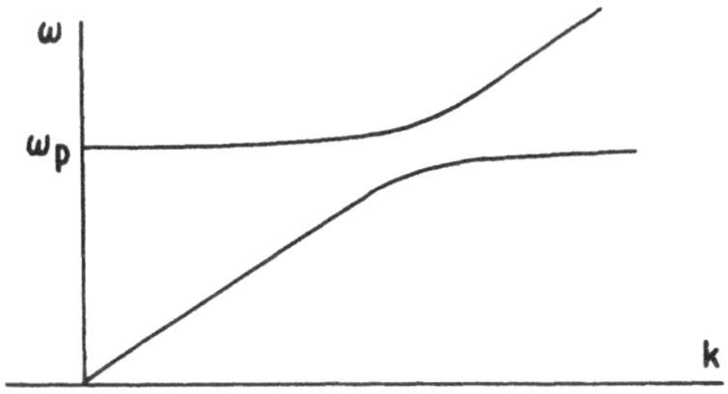

Fig. (I.4.2) Sketch of ω vs. k curves for the model dispersion relation of Eq. (I.4.14).

Similarly, we can show that the coupling between two negative-energy waves introduces no instability.

We now briefly examine the beam-plasma instability. Here, we let a cold electron beam streaming through a background cold plasma. Let the beam velocity be v_b and density n_b. For the background plasma, we have density n_o and $m_i \to \infty$, i.e., immobile ions. For simplicity of analysis, we assume the beam is weak; $n_b \ll n_o$. The corresponding electrostatic dispersion relation can then be shown to be

$$\varepsilon_h = 1 - \frac{\omega_{pe}^2}{\omega^2} - \frac{\omega_{pb}^2}{(\omega-kv_b)^2} = 0. \qquad (I.4.15)$$

Equation (I.4.15) can also be written as

$$\omega_1^2 = \omega_{pe}^2/[1 - \omega_{pb}^2/(\omega_1-kv_b)^2] , \qquad (I.4.16)$$

and

$$(\omega_2 - kv_b)^2 = \omega_{pb}^2/(1-\omega_{pe}^2/\omega_2^2) . \qquad (I.4.17)$$

For $|kv_b| < \omega_{pe}$, we have

$$\omega_1 \approx \pm \omega_{pe}/[1 - \omega_{pb}^2/\omega_{pe}^2]^{1/2} \simeq \pm \omega_{pe} , \qquad (I.4.18)$$

and

$$\omega_2 = kv_b \pm i(\omega_{pb}/\omega_{pe})kv_b . \qquad (I.4.19)$$

Meanwhile, for $|kv_b| > \omega_{pe}$, we have

$$\omega_1 \simeq \pm \omega_{pe}/[1 - \omega_{pb}^2/k^2v_b^2]^{1/2} \simeq \pm \omega_{pe} , \qquad (I.4.20)$$

and

$$\omega_2 \simeq kv_b \pm \omega_{pb} . \qquad (I.4.21)$$

The corresponding ω-k dispersion curves are sketched in Fig. (I.4.3).

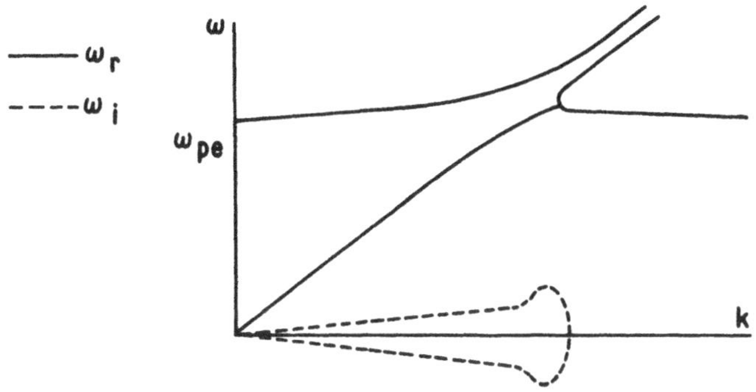

Fig. (I.4.3) Sketch of ω vs. k cruves for the beam-plasma instability dispersion relation of Eq. (I.4.15).

By the analogy with the model problem, we can understand this instability in terms of the coupling between the positive-energy wave with $\omega \simeq \omega_{pe}$ and the negative-energy wave with $\omega \simeq kv_b - \omega_{pb}$. This is clear by noting the respective dielectric constants are

$$\varepsilon_{h1} = 1 - \omega_{pe}^2/\omega_1^2 \ , \tag{I.4.22}$$

and

$$\varepsilon_{h2} = 1 - \omega_{pb}^2/(\omega_2 - kv_b)^2 \ . \tag{I.4.23}$$

§I.5 Nyquist technique for stability analysis

In many practical applications, the dispersion relation $\varepsilon(\omega,\underset{\sim}{k}) = 0$ is generally too complicated to solve analytically for instability growth rates. In fact, the first thing one would like to know is if the dispersion relation admits solutions of ω with Im $\omega > 0$. For this purpose, there exists the powerful Nyquist technique.

Let us assume that ε be analytic in the Im $\omega > 0$ half plane and that it has a finite number of zeroes at $\omega = \omega_m$ for $m = 1,\dots,N$. That is, ω_m's are the unstable solutions. We now construct the following function

$$G(\omega) = \frac{1}{\varepsilon(\omega)} \frac{\partial \varepsilon}{\partial \omega} \quad . \tag{I.5.1}$$

It is clear that $G(\omega)$ has poles at $\omega = \omega_m$. We now define a contour c in the $\mathrm{Im}\,\omega > 0$ plane to be $c = c_1 + c_2$ where c_1 lies above the real ω axis and c_2 is an infinite semicircle [c.f., Fig. (I.5.1)].

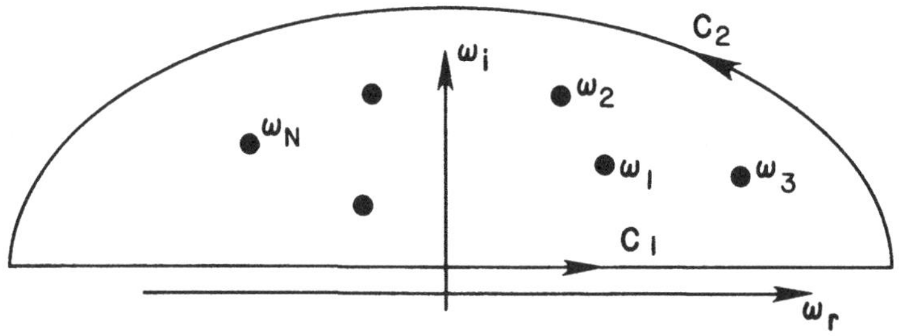

Fig. (I.5.1) Integration contour in the complex ω plane for the Nyquist stability analysis.

From Cauchy theorem, we find

$$\frac{1}{2\pi i} \int_c G(\omega) d\omega = \sum_{m=1}^{N} \mathrm{Res}\, G(\omega=\omega_m) = \sum_{m=1}^{N} \mathrm{Res}\, \left((1/\varepsilon)(\partial\varepsilon/\partial\omega)\right)_{\omega=\omega_m} \quad . \tag{I.5.2}$$

Now, near $\omega = \omega_m$, we have

$$\varepsilon(\omega) \simeq d_1(\omega-\omega_m)^{P_m} + d_2(\omega-\omega_m)^{P_m+1} + \dots \quad , \tag{I.5.3}$$

or

$$G(\omega) \simeq \frac{P_m}{(\omega-\omega_m)} \quad . \tag{I.5.4}$$

Equation (I.5.2) then reduces to

$$\frac{1}{2\pi i}\int_c G(\omega)d\omega = \sum_{m=1}^{N} P_m \ . \qquad (I.5.5)$$

Equation (I.5.5) shows that given $\varepsilon(\omega,\underset{\sim}{k})$ being analytic in the Im $\omega > 0$ plane, there then exists instabilities if

$$\frac{1}{2\pi i}\int_c \left(\frac{1}{\varepsilon}\frac{\partial\varepsilon}{\partial\omega}\right)d\omega = M = \text{a positive integer.} \qquad (I.5.6)$$

This is the, sometimes, called Nyquist stability theorem.

We now further explore Eq. (I.5.6). First, we note that

$$\frac{1}{2\pi i}\int_c \left(\frac{1}{\varepsilon}\frac{\partial\varepsilon}{\partial\omega}\right)d\omega = \frac{1}{2\pi i}\,\ell n\left[\frac{\varepsilon(c_e)}{\varepsilon(c_s)}\right] \ . \qquad (I.5.7)$$

Now, let

$$\varepsilon(c_e) = \varepsilon(c_s)\exp(i2\pi M) \quad , \qquad (I.5.8)$$

we, again, recover the result of Eq. (I.5.6). Here, however, M corresponds to the number of times that the $\varepsilon(\omega)$ curve when mapped along the contour c encircles the origin of the ε plane in the counter-clockwise direction. Thus, we reduce the stability analysis to the problem of mapping $\varepsilon(\omega)$ along the c contour.

Let us illustrate this technique with the beam-plasma instability discussed in Sec. §I.4. The corresponding linear dispersion relation is

$$\varepsilon_h(\omega) = 1 - \frac{\omega_p^2}{\omega^2} - \frac{\omega_b^2}{(\omega-kv_b)^2} = 0 \ , \qquad (I.4.15)$$

with $|\omega_p| >> |\omega_b|$ assumed for simplicity. First, we consider the $kv_b < \omega_p$ limit which we know is unstable. Now since ε_h has poles at $\omega = 0$ and $\omega = kv_b$. The c_1 contour needs to be deformed about the poles as shown in Fig. (I.5.2).

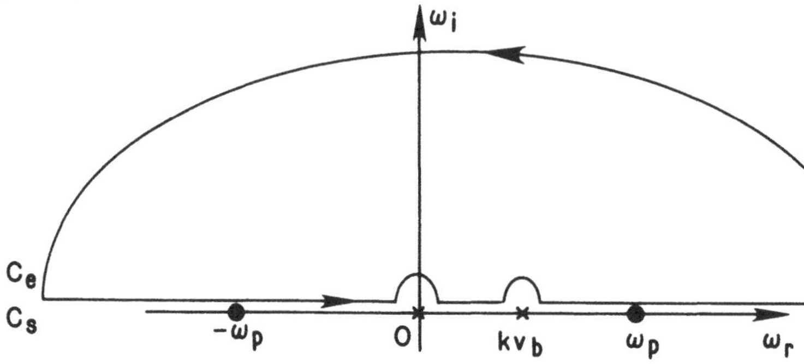

Fig. (I.5.2) ω-plane contour for the Nyquist analysis of the beam-plasma instability dispersion relation of Eq. (I.4.15) in the $kv_b < \omega_p$ limit.

In Fig. (I.5.2), the radius of the two (small) semicircles about the poles, δ, is taken to be infinitesimally small but finite. The mapping of $\varepsilon_h(\omega)$ along $c_\omega = c_1 + c_2$ is shown in Fig. (I.5.3).

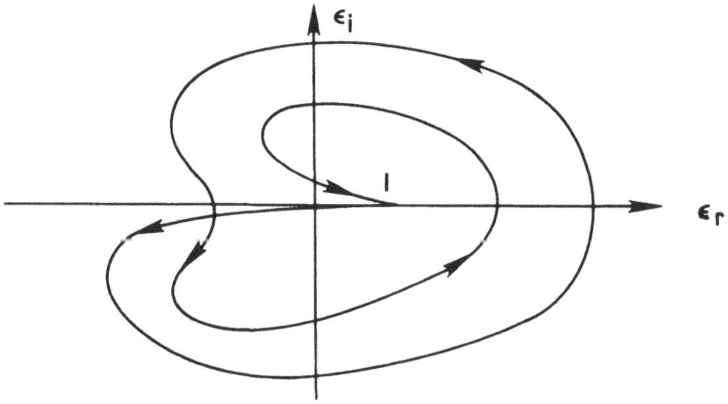

Fig. (I.5.3) Mapping of ε_h given by Eq. (I.4.15) along the ω-plane contour shown in Fig. (I.5.2).

Thus, $\epsilon_h(\omega)$ along c_ω encircles the origin once in the counter-clockwise direction. That is, M = 1 and we have one unstable solution. This, of course, agrees with the results obtained in Sec. §I.4. Next, we consider the $k_o v_b > \omega_p$ limit, which we know is stable. The corresponding c_ω contour is given in Fig. (I.5.4).

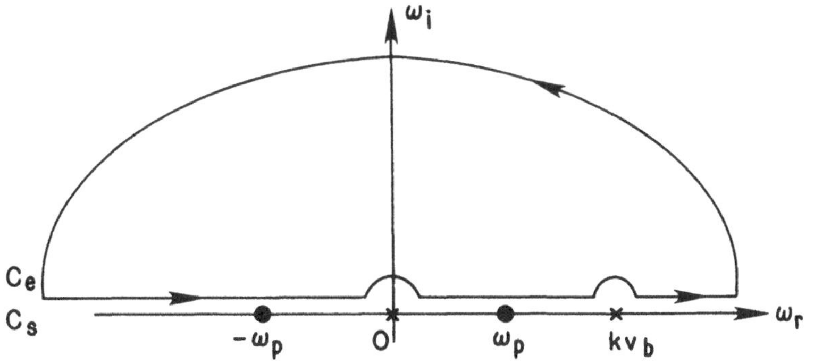

Fig. (I.5.4) ω-plane contour for the Nyquist analysis of the beam-plasma instability dispersion relation of Eq. (I.4.15) in the $kv_b > \omega_p$ limit.

The resultant mapping of $\epsilon_h(\omega)$ is shown in Fig. (I.5.5).

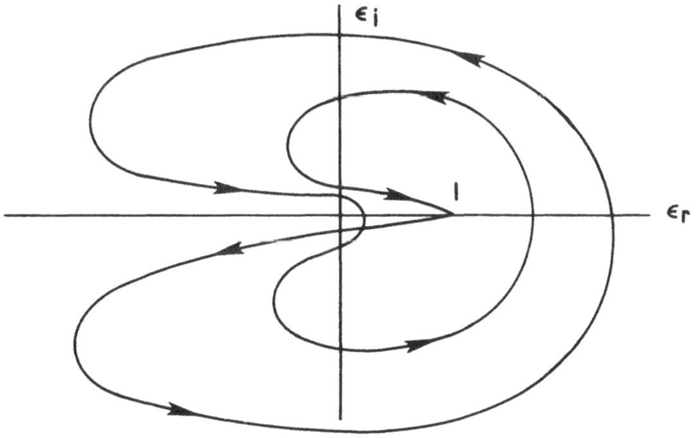

Fig. (I.5.5) Mapping of ϵ_h given by Eq. (I.4.15) along the ω-plane contour shown in Fig. (I.5.4).

Thus, in this case $\varepsilon_h(\omega)$ mapped along c_ω does not encircle the origin and we have M = 0; i.e., we have no instability as it should.

Now the Nyquist technique for stability analyses is not limited to simple dispersion relation. As an example, we shall apply it to a stability analysis governed by a differential equation. This application, which is relevant to eigenmode stability problems to be discussed later, is presented here to further demonstrate the Nyquist technique. Let us consider the following model differential equation.

$$[\frac{d^2}{dx^2} + \omega^2 + i\omega\nu - \omega_o^2 - x^2]\delta\phi(x) = 0 \ , \qquad \text{(I.5.9)}$$

where $\nu > 0$ and the boundary conditions are

$$|\delta\phi(x)| \to 0 \quad \text{as} \quad |x| \to \infty \ . \qquad \text{(I.5.10)}$$

As a physical motivation, Eq. (I.5.9) may be regarded as modelling the wave equation for Bohm - Gross (warm electron plasma) waves in a Gaussian density cavity. In order to apply the Nyquist technique, we need to obtain an equivalent dispersion relation in ω. For that purpose, we apply the operation $\int_{-\infty}^{\infty} dx\ \delta\phi^* (x)$ to Eq. (I.5.9). Noting Eq. (I.5.10), we find

$$\varepsilon(\omega) \equiv \omega^2 + i\ \omega\nu - \omega_o^2 - [\langle|x\delta\phi|^2\rangle + \langle|d\delta\phi/dx|^2\rangle]/\langle|\delta\phi|^2\rangle = 0, \quad \text{(I.5.11)}$$

where

$$\langle A\rangle \equiv \int_{-\infty}^{\infty} dx\ A. \qquad \text{(I.5.12)}$$

Now since $\varepsilon(\omega)$ diverges as $|\omega| \to \infty$, we shall choose the radius of the semi-circle contour, R, to be arbitrary large but finite [refer to Fig. (I.5.6)].

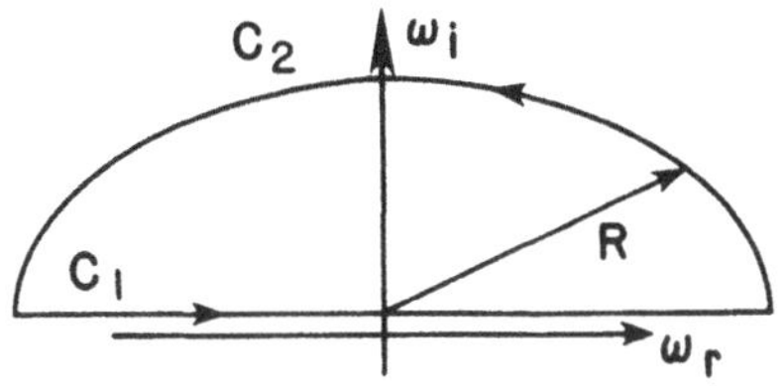

Fig. (I.5.6) ω-plane contour for the Nyquist analysis of the "dispersion relation" of Eq. (I.5.11).

The mapping of $\varepsilon(\omega)$ along c_ω is shown in Fig. (I.5.7); from which we conclude that Eq. (I.5.9) predicts no instability.

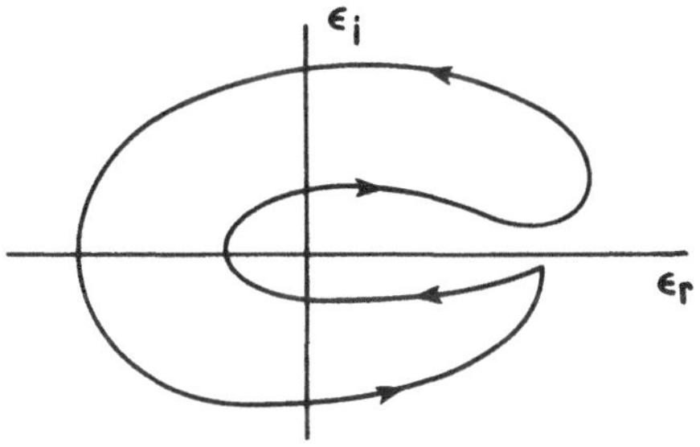

Fig. (I.5.7) Mapping of ε given by Eq. (I.5.11) along the ω-plane contour shown in Fig. (I.5.6).

In fact, we may choose c_1 to be infinitesimally below the ω_r axis (i.e., Im $c_1 \to 0^-$) and prove that no marginally stable solution exists either. Finally, we remark that, while in this model example the above conclusions can also be obtained by either noting that Eq. (I.5.9) is a Weber equation or solving Eq. (I.5.11) which is a quadratic equation in ω, the Nyquist technique is more powerful and applicable to a broader range of problems.

Homework # 3

(1) The following dispersion relation describes beam-plasma interaction including collisional effects on the background electrons.

$$\varepsilon = 1 - \frac{\omega_p^2}{\omega(\omega+i\nu)} - \frac{\omega_b^2}{(\omega-kv_b)^2} = 0 \text{ with } \nu > 0. \tag{H.3.1}$$

Show that in the $|kv_b| \gg \omega_p$ limit, which is stable in the reactive limit (i.e., $\nu = 0$), the negative-energy wave, (i.e., the slow beam mode) is unstable due to the positive collisional dissipation.

(2) Consider electrons streaming through background immobile ions with velocity v_b. The corresponding electrostatic dispersion relation is

$$\varepsilon = 1 - \frac{\omega_b^2}{(\omega-kv_b)^2} = 0 \tag{H.3.2}$$

Suppose there is an electric field of the plane-wave form

$$\delta \underset{\sim}{E}(x,t) = \frac{1}{2}\left[\delta\hat{E}\exp(-i\omega t + ikx) + \text{c.c.}\right] ,$$

where $\delta\hat{E}$ is independent of t and x. Calculate the phase-averaged particle kinetic energy density for both the slow, $\omega = kv_b - \omega_b$, and the fast modes.

(3) Using Nyquist technique to prove that, as discussed in Sec. §I.4, the model dispersion relation

$$\varepsilon = 1 - \frac{\omega_p}{\omega} + \frac{\omega_b}{\omega-kv_o} = 0 \tag{I.4.7}$$

predicts

(3.a) stability for $kv_o \ll \omega_p$

(3.b) stability for $kv_o \gg \omega_p$

(3.c) instability for $kv_o = \omega_p$.

§I.6 Absolute and convective instabilities

[Ref. Ch. 2 in R. J. Brigg's "Electron Stream Interaction with Plasmas" M.I.T. Press.]

Up to now, our definition of instabilities (Sec. §I.4) is based on perturbations which have plane-wave spatial dependence; i.e., $\exp(i\underset{\sim}{k}\cdot\underset{\sim}{x})$. In other words, the perturbations are taken to be periodic in $\underset{\sim}{x}$ and, hence, have infinite spatial extent. In reality, however, the perturbations are generally of finite spatial extent. Treating as an initial-value problem, there then exists two types of time-asymptotic ($t \to \infty$) behaviors at a fixed finite spatial point. One is the absolute instability characterizing by perturbations which become unbounded as $t \to \infty$ at every fixed finite point in space. The other is the convective instability characterizing by perturbations which propagate and grow along the system such that, at any fixed finite point in space, the perturbations vanish time-asymptotically. A sketch of these two-types of time-asymptotic behaviors is shown in Fig. (I.6.1).

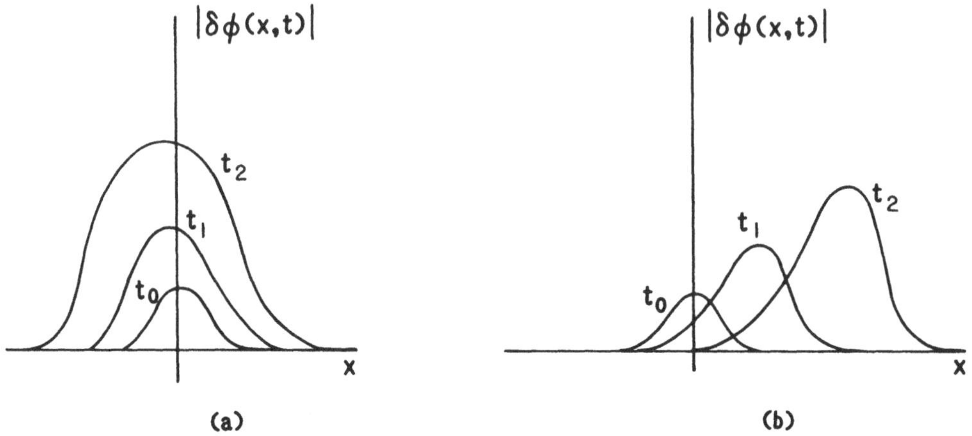

Fig. (I.6.1) Sketches of time-asymptotic perturbations of (a) an absolute instability where $|\delta\phi(x,t)| \to \infty$ as $t \to \infty$ and (b) a convective instability where $|\delta\phi(x,t)| \to 0$ as $t \to \infty$.

We emphasize that the differentiation of these two types of instability is not simply an academic exercise but carries important practical implications. This is because (i) instabilities are grown out of low-level thermal fluctuations and (ii) the unstable region is, in practice, of a limited spatial extent. For convective instabilities, the perturbations are spatially amplified inside the unstable region. Thus, their amplitudes as well as their nonlinear consequences can be estimated. For example, if the unstable region is small, the instabilities may be practically insignificant. On the other hand, absolute instabilities can grow to large amplitudes such that their saturation can only be due to nonlinear effects.

In terms of operators, let us assume the perturbation, $\delta\phi(t,x)$ is governed by the following wave equation

$$\varepsilon\left(i\,\frac{\partial}{\partial t}\,,\;-i\,\frac{\partial}{\partial x}\right)\delta\phi(t,x) = 0, \qquad \text{(I.6.1)}$$

subject to an initial perturbation, $\delta\phi(t=0,x)$. Here, we reiterate that the system is linearly unstable; i.e., the linear dispersion relation

$$\varepsilon(\omega,k) = 0 \ , \tag{I.6.2}$$

has unstable roots, Im $\omega > 0$, for real k. We, furthermore, consider the one-dimensional case only. Generalization to higher dimensionalities is straightforward.

Laplace transforming Eq. (I.6.1) in time, we obtain

$$\varepsilon(\omega, -i \frac{\partial}{\partial x})\delta\phi(\omega,x) = s(x), \tag{I.6.2}$$

where s(x) represents the effective initial perturbations. Let us denote the Green's function response be $\delta G(\omega,x)$ such that

$$\varepsilon(\omega, -i \frac{\partial}{\partial x})\delta G(\omega,x) = \delta(x) \quad . \tag{I.6.3}$$

We then have

$$\delta\phi(\omega,x) = \int dx' s(x')\delta G(\omega,x-x'). \tag{I.6.4}$$

Equation (I.6.4) shows that it is sufficient to examine Eq. (I.6.3) in order to understand the time-asymptotic behaviors.

Formulating in terms of the Green's function, we have

$$\delta G(t,x) = \int_{c_\omega} \frac{d\omega}{2\pi} \delta G(\omega,x) \ e^{-i\omega t}, \tag{I.6.5}$$

and

$$\delta G(\omega,x) = \int_{-\infty}^{\infty} \frac{dk}{2\pi} \frac{1}{\varepsilon(\omega,k)} e^{ikx} . \qquad (I.6.6)$$

Here, the ω integration contour, c_ω, must lie above the zeroes of $\varepsilon(\omega,k)$ for a given real k in order to ensure the causality condition is satisfied. Furthermore, as being physically resonable, we assume that $\varepsilon(\omega,k)$ is sufficiently well-behaved as $|\omega|$ or $|k|$ approaches infinity. The ω and k integration contours for, respectively, $t > 0$ and $x > 0$ can then be closed as shown in Fig. (I.6.2).

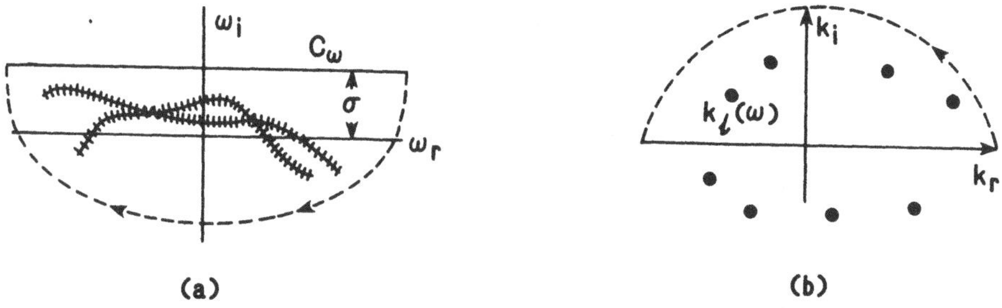

Fig. (I.6.2) Sketches of contours for (a) ω-plane integration of Eq. (I.6.5) and (b) k-plane integrations of Eq. (I.6.6).

In Fig. (I.6.2a), ┼┼┼┼┼ corresponds to the solutions of the dispersion relation $\varepsilon(\omega_j, k) = 0$ for real k and $j = 1,\ldots, N$ corresponds to the jth branch of the normal modes. That Im $\omega_j > 0$ for some real k, of course, is consistent with the fact that the system is unstable. Further, as discussed above, Im $c_\omega = \sigma >$ Max Im (ω_j) and, hence, $\delta G(\omega,x)$ is analytic for ω on and above c_ω. Meanwhile, in Fig. (I.6.2b), • corresponds to solutions of the dispersion relation for a <u>given ω on c_ω</u>; i.e., $\varepsilon[\omega,k_\ell(\omega)] = 0$, $\ell = 1,\ldots M$ and ω on c_ω.

Since $\delta G(\omega,x)$ is analytic only for $\omega_i \geq \sigma$, we need to analytically continuate it as we close the ω-integration contour in the $\omega_i < 0$ half plane. Referring to Fig. (I.6.3a), let us take one point on c_ω, $\omega = \omega_0$ and continuate it downward.

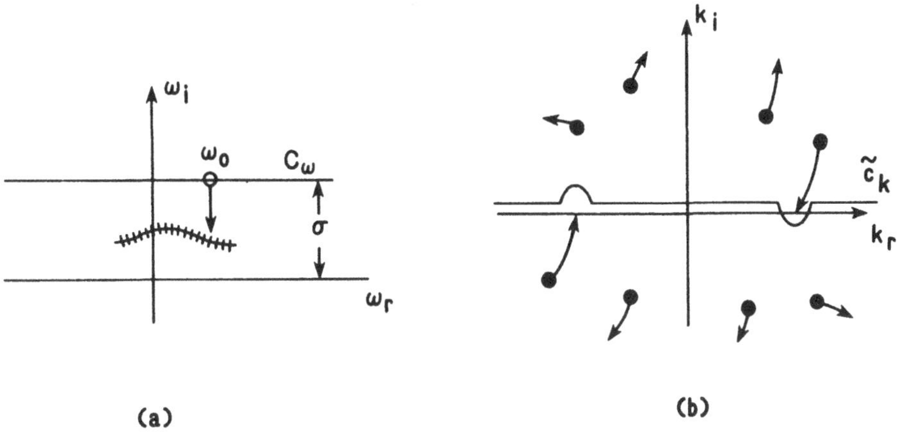

Fig. (I.6.3) Deformed contours for analytic continuations of integrations in (a) the ω plane and (b) the k plane.

Since $k_\ell = k_\ell(\omega)$ depends on ω, the poles in k plane will move around as ω_0 moves downward. In particular, we know that, as $\omega_0 \to \omega_j$, some pole(s) in k plane must move toward the k_r-axis [c.f. Fig. (I.6.3b)]. This is because ω_j is defined by $\varepsilon(\omega_j, k = \text{real}) = 0$. As, say, $k_\ell(\omega_s)$ becomes real, $\delta G(\omega,x)$ as defined by Eq. (I.6.6) becomes non-analytic since $k_\ell(\omega_j)$ lies on the k-integration contour c_k which is the k_r axis. In this respect, $\omega_j(k)$'s are the branch lines of $\delta G(\omega,x)$ in the ω plane. In order to analytically continuate $\delta G(\omega,x)$ below $\omega_j(k)$, it is then necessary to deform the k-integration contour around those poles which approach or cross the k_r axis; i.e., we must choose a new contour, $\tilde{c}_k$, as illustrated in Fig. (I.6.3b). The analytic continuation of δG as defined by Eq. (I.6.6) is then given by

$$\delta\hat{G}(\omega,x) = \int_{\tilde{c}_k} \frac{dk}{2\pi} \frac{1}{\varepsilon(\omega,k)} \cdot e^{ikx} \quad . \tag{I.6.7}$$

Thus, with $\tilde{c}_k$ so chosen, the k plane poles, $k_\ell(\omega)$, never cross the $\tilde{c}_k$ as ω moves downward. That is, when we close the k-plane integration in the $k_i > 0$ half plane, the <u>identities</u> of those poles which contribute to the integral do <u>not</u> change. We then have

$$\hat{\delta G}(\omega,x) = i \sum_{\ell} \frac{e^{ik_{\ell}(\omega)}}{\partial\varepsilon(\omega,k_{\ell})/\partial k_{\ell}} , \tag{I.6.8}$$

with the understanding that only those k_ℓ which have Im $k_\ell > 0$ when Im $\omega \geq \sigma$ are included.

Substituting Eq. (I.6.8) into Eq. (I.6.5), it is clear that poles of $\hat{\delta G}$ will contribute to the ω integration. In particular, in the time-asymptotic limit ($t \to \infty$), the pole with maximum ω_i dominates. Let this ω-plane pole be ω_s. Thus, $\partial\varepsilon[\omega_s, k_s(\omega_s)]/\partial k_s = 0$; i.e., at $\omega = \omega_s$, $\varepsilon(\omega_s, k) = 0$ has a double root at $k = k_s$. In other words, as $\omega \to \omega_s$, we have two k-plane poles converging at $k = k_s$. Expanding $\varepsilon(\omega,k)$ about ω_s and k_s, respectively, we find

$$\varepsilon(\omega,k) \simeq \left(\frac{\partial\varepsilon}{\partial\omega_s}\right)(\omega-\omega_s) + \frac{1}{2}\left(\frac{\partial^2\varepsilon}{\partial k_s^2}\right)(k-k_s)^2. \tag{I.6.9}$$

Substituting Eq. (I.6.9) into Eq. (I.6.7), it is fairly straightforward to show that $\hat{\delta G}(\omega_s,x)$ is singular (regular) if the two converging poles lie on the opposite (same) side of $\tilde{c}_k$. In particular, we have, for the singular case, that, noting

$$\lim_{|\alpha| \to 0^+} \int_{-\infty}^{\infty} \frac{dx}{(x+i\alpha)(x-i\alpha)} = \frac{\pi}{\alpha} \text{ for Re } \alpha > 0 , \tag{I.6.10}$$

$$\hat{\delta G}(\omega,x) \simeq \left[2 \left(\frac{\partial\varepsilon}{\partial\omega_s}\right)\left(\frac{\partial^2\varepsilon}{\partial k_s^2}\right)\right]^{-1/2} (\omega-\omega_s)^{-1/2} e^{ik_s x} \tag{I.6.11}$$

for $\omega \simeq \omega_s$. Equation (I.6.11) into Eq. (I.6.5), we find that

$$\delta G(t,x) \simeq \left[2\left(\frac{\partial\varepsilon}{\partial\omega_s}\right)\left(\frac{\partial^2\varepsilon}{\partial k_s^2}\right)\right]^{-1/2} e^{-i\omega_s t+ik_s x} \int_{c_\omega} \frac{d\omega}{2\pi} \cdot \frac{e^{-i(\omega-\omega_s)t}}{(\omega-\omega_s)^{1/2}} , \tag{I.6.12}$$

where $\tilde{c}_\omega$ as sketched in Fig. (I.6.4) must go around the branch cut.

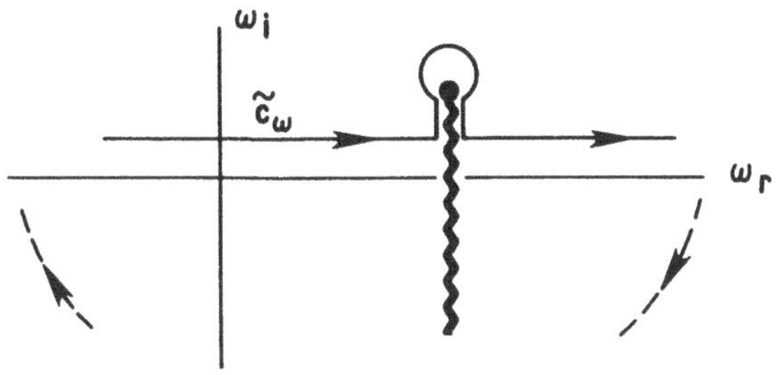

Fig. (I.6.4) Sketch of the integration contour around the branch cut defined by Eq. (I.6.12).

It is easy to show that

$$\lim_{t \to \infty} \left| \int_{\tilde{c}_\omega} \frac{d\omega}{2\pi} \frac{e^{-i(\omega-\omega_s)t}}{(\omega-\omega_s)^{1/2}} \right| \propto \frac{1}{t^{1/2}} . \qquad (I.6.13)$$

Thus, we obtain, in the time asymptotic limit,

$$\lim_{t \to \infty} \delta G(t,x) \propto t^{-1/2} \exp[i(k_s x - i\omega_s t)]. \qquad (I.6.14)$$

Equation (I.6.14) clearly shows that if Im $\omega_s > 0$ we then have $|\delta G(t,x)|$ or $|\delta\phi(t,x)| \to \infty$ as $t \to \infty$, i.e., an absolute instability. Otherwise, for Im $\omega_s \leq 0$, we have $|\delta G|$ or $|\delta\phi| \to 0$ as $t \to \infty$; i.e., a convective instability.

Let us summarize the results obtained in this section. First, we assume the system is linearly unstable, i.e., the linear dispersion relation $\varepsilon(\omega,k) = 0$ has unstable (Im $\omega > 0$) roots for real k. We then show that, if (i) there exists two k solutions of $\varepsilon(\omega,k) = 0$ which lie, respectively, on the upper and lower half k plane for Im $\omega \geq \sigma \equiv \mathrm{Im}(c_\omega)$, and (ii) the two k solutions coalesce ($\partial\varepsilon/\partial k_s = 0$) at $\omega = \omega_s$ with Im $\omega_s > 0$, we then have an absolute instability. Otherwise, the system only exhibits convective instability. We remark that the condition $\partial\varepsilon/\partial k_s = 0$ at $\omega = \omega_s$ is equivalent to the condition that the group velocity $\partial\omega_s/\partial k_s = 0$. Thus, the above requirements for absolute instabilities are also consistent with our physical intuition that absolute instabilities correspond to standing-still wavepackets which, while growing in time, also tend to spread out in both positive and negative directions. Finally, we emphasize that the nature of the instability depends crucially on the reference frame. This is intuitively expected, because for an observer travelling with the group velocity of the wave packet a convective instability in the laboratory frame will appear to be an absolute instability to him/her.

Let us illustrate the application of the above stability criteria with the following three examples:

(i) Consider the model dispersion relation

$$(k-\omega/v_1)(k-\omega/v_2) = -k_o^2 , \tag{I.6.15}$$

where $v_1 v_2 > 0$. By simply considering the $k \to 0$ limit, it is clear that Eq. (I.6.15) predicts instability. Meanwhile, in the limit Im $c_\omega \to \infty$ we find $k_1 \to \omega/v_1$ and $k_2 \to \omega/v_2$; i.e., the two k-plane poles lie on the same side of k-integration contour c_k and, hence, the instability is convective.

(ii) We now consider an another model dispersion relation

$$(k-\omega/v_1)(k+\omega/v_2) = k_o^2 , \tag{I.6.16}$$

here, again, $v_1 v_2 > 0$ and the system is unstable. Now as Im $c_\omega \to \infty$, $k_1 \to \omega/v_1$ and $k_2 \to -\omega/v_2$ and, thus, two k-plane poles are on the opposite side of c_k. We now calculate ω_s where k_1 and k_2 coalesce. From, $\partial\omega_s/\partial k_s = 0$, we find

$$k_s = \frac{\omega_s}{2} (1/v_1 - 1/v_2). \qquad (I.6.17)$$

For simplicity, we further assume $v_1 = v_2$. Then $k_s = 0$ and, from Eq. (I.6.16), we find

$$\omega_s^2 = - k_o^2 v_1^2 \quad \text{or} \quad \omega_s = \pm\, i k_o v_1 . \qquad (I.6.18)$$

Thus, Im $\omega_s > 0$ and we have an absolute instability.

(iii) Finally, we consider the beam-plasma instability given by the dispersion relation

$$1 - \frac{\omega_p^2}{\omega^2} - \frac{\omega_b^2}{(\omega - k v_b)^2} = 0. \qquad (I.6.19)$$

Now, as in Example (i), we have, in the limit $c_\omega \to \infty$, $k_{1,2} \to \omega/v_b$ and, thus, this instability is convective.

CHAPTER II

Linear Waves and Instabilities in Uniform Unmagnetized Plasmas

In the previous chapter, we have discussed some general wave properties in a uniform dielectric medium; assuming that the dielectric tensor, $\underset{\approx}{D}(\omega,\underset{\sim}{k})$, is given. In this and the next chapters, we shall derive specific expressions for $\underset{\approx}{D}$ and discuss the results. Since only high-temperature plasmas are of interest here, we assume, unless otherwise stated, that the plasmas are collisionless. Thus, the plasma dynamics obey the following Vlasov equation

$$\left[\frac{\partial}{\partial t} + \underset{\sim}{v} \cdot \frac{\partial}{\partial \underset{\sim}{x}} + \frac{q}{m}\left(\underset{\sim}{E} + \frac{\underset{\sim}{v}\times\underset{\sim}{B}}{c}\right) \cdot \frac{\partial}{\partial \underset{\sim}{v}}\right] f(\underset{\sim}{x},\underset{\sim}{v},t) = 0,$$

plus the Maxwell's equations.

§II.1 Solution of the linearized Vlasov equation

Let

$$f = f_0(\underset{\sim}{x},\underset{\sim}{v}) + \delta f(\underset{\sim}{x},\underset{\sim}{v},t) \quad , \qquad \text{(II.1.1)}$$

where f_0 and δf are, respectively, the equilibrium and perturbed velocity distribution functions. Now with $\underset{\sim}{B}_0=\underset{\sim}{E}_0=0$, we find $\partial f_0/\partial \underset{\sim}{x} = 0$, i.e.,

$$f_0 = f_0(\underset{\sim}{v}) \quad . \qquad \text{(II.1.2)}$$

As to δf, we shall assume, for simplicity of discussion, that the wave perturbations are electrostatic; i.e., $\delta\underset{\sim}{B} \propto \underset{\sim}{\nabla} \times \delta\underset{\sim}{E} = 0$. Thus, we have

$$(\frac{\partial}{\partial t} + \underset{\sim}{v} \cdot \frac{\partial}{\partial \underset{\sim}{x}})\delta f = - \frac{q}{m}\, \delta \underset{\sim}{E} \cdot \frac{\partial}{\partial \underset{\sim}{v}}(f_o + \delta f) \quad . \tag{II.1.3}$$

Linearizing Eq. (II.1.3) or, specifically, assuming

$$\frac{|\partial \delta f/\partial \underset{\sim}{v}|}{|\partial f_o/\partial \underset{\sim}{v}|} << 1 \quad , \tag{II.1.4}$$

we arrive at the linearized Vlasov equation

$$\pounds_o \delta f_1 \equiv (\frac{\partial}{\partial t} + \underset{\sim}{v} \cdot \frac{\partial}{\partial \underset{\sim}{x}})\delta f_1 = - \frac{q}{m}\, \delta \underset{\sim}{E} \cdot \frac{\partial f_o}{\partial \underset{\sim}{v}} \quad . \tag{II.1.5}$$

Equation (II.1.5) is a first-order partial differential equation and, hence, can be readily solved by integrating along its characteristics or, in this case, the zeroth-order (unperturbed) particle orbits

$$\frac{d\underset{\sim}{x}'(t')}{dt'} = \underset{\sim}{v}'(t') \tag{II.1.6}$$

and

$$\frac{d\underset{\sim}{v}'(t')}{dt'} = 0 \quad ; \tag{II.1.7}$$

with the boundary conditions that

$$\underset{\sim}{x}'(t' = t) = \underset{\sim}{x} \tag{II.1.8}$$

and

$$\underset{\sim}{v}'(t' = t) = \underset{\sim}{v} \quad . \tag{II.1.9}$$

The solution to Eq. (II.1.5) is then

$$\delta f_1(\underset{\sim}{x},\underset{\sim}{v},t) = \delta f_1(\underset{\sim}{x}_o,\underset{\sim}{v}_o,t=0) - \frac{q}{m}\int_0^t dt'\delta\underset{\sim}{E}(\underset{\sim}{x}',t')\cdot\frac{\partial f_o(\underset{\sim}{v}')}{\partial\underset{\sim}{v}'} \quad . \tag{II.1.10}$$

Here, $\underset{\sim}{x}_o = \underset{\sim}{x}'(t'=0)$ and $\underset{\sim}{v}_o = \underset{\sim}{v}'(t'=0)$. Noting that $\underset{\sim}{v}' = \underset{\sim}{v}$ and, therefore, $\underset{\sim}{x}'(t') = \underset{\sim}{x} - \underset{\sim}{v}(t-t')$, Eq. (II.1.10) becomes

$$\delta f_1(\underset{\sim}{x},\underset{\sim}{v},t) = \delta f_1(\underset{\sim}{x}-\underset{\sim}{v}t,\underset{\sim}{v},t=0) - \frac{q}{m}\frac{\partial f_o(\underset{\sim}{v})}{\partial\underset{\sim}{v}}\cdot\int_0^t dt'\ \delta\underset{\sim}{E}(\underset{\sim}{x}',t') \quad . \tag{II.1.11}$$

Substituting Eq. (II.1.11) into the Poisson's equation

$$\underset{\sim}{\nabla}\cdot\delta\underset{\sim}{E} = 4\pi\sum_j q_j\int\delta f_{1j}\,d^3\underset{\sim}{v}\ ;\ j = \text{species}, \tag{II.1.12}$$

we then have the linear dynamics completely determined.

We now examine the linearization from a different point of view. We first note that the full perturbed equation can be also written as

$$\pounds_v\delta f = \left[\frac{\partial}{\partial t} + \underset{\sim}{v}\cdot\frac{\partial}{\partial\underset{\sim}{x}} + \frac{q}{m}\delta\underset{\sim}{E}\cdot\frac{\partial}{\partial\underset{\sim}{v}}\right]\delta f = -\frac{q}{m}\delta\underset{\sim}{E}\cdot\frac{\partial f_o}{\partial\underset{\sim}{v}} \quad . \tag{II.1.13}$$

δf, thus, can be formally solved as

$$\delta f(\underset{\sim}{x},\underset{\sim}{v},t) = \delta f(\underset{\sim}{X}_o,\underset{\sim}{V}_o,t=0) - \frac{q}{m}\int_0^t dt'\delta\underset{\sim}{E}(\underset{\sim}{X}',t')\cdot\frac{\partial f_o}{\partial\underset{\sim}{V}'}(\underset{\sim}{V}')\ ; \tag{II.1.14}$$

where $\underset{\sim}{X}'$ and $\underset{\sim}{V}'$ are, however, particle's <u>exact</u> phase-space trajectories; i.e.,

$$\frac{d\underset{\sim}{X}'}{dt'} = \underset{\sim}{V}' \text{ and } \frac{d\underset{\sim}{V}'}{dt'} = \frac{q}{m}\delta\underset{\sim}{E}(\underset{\sim}{X}',t') \quad ; \tag{II.1.15}$$

with boundary conditions similar to those given by Eqs. (II.1.8) and (II.1.9). Now if we let

$$\underset{\sim}{X}' = \underset{\sim}{x}' + \delta\underset{\sim}{x}' \tag{II.1.16}$$

and

$$\underset{\sim}{V}' = \underset{\sim}{v}' + \delta\underset{\sim}{v}' \quad , \tag{II.1.17}$$

where $\underset{\sim}{x}'$ and $\underset{\sim}{v}'$ are the unperturbed orbits, Eqs. (II.1.6) and (II.1.7), we then have

$$\frac{d\delta\underset{\sim}{x}'}{dt} = \delta\underset{\sim}{v}' \quad \text{and} \quad \frac{d\delta\underset{\sim}{v}'}{dt} = \frac{q}{m}\,\delta\underset{\sim}{E}(\underset{\sim}{X}',t') \quad . \tag{II.1.18}$$

Substituting Eqs. (II.1.16) and (II.1.17) into Eq. (II.1.14) and expanding $\underset{\sim}{X}'$ and $\underset{\sim}{V}'$ about, respectively, $\underset{\sim}{x}'$ and $\underset{\sim}{v}'$ it is then clear by comparing the results with Eq. (II.1.10) that linearization corresponds to assuming

$$\left|\int_0^t \left[\delta\underset{\sim}{x}' \cdot \frac{\partial}{\partial\underset{\sim}{x}'}\,\delta\underset{\sim}{E}(\underset{\sim}{x}',t')\right]dt'\right| \ll \left|\int_0^t \delta\underset{\sim}{E}(\underset{\sim}{x}',t')dt'\right| \tag{II.1.19}$$

and

$$\left|\int_0^t dt'\;\delta E_\ell(\underset{\sim}{x}',t')\delta v'_m\right| \left|\frac{\partial^2 f_o}{\partial v_m \partial v_\ell}\right| \ll \left|\int_0^t dt'\;\delta E_\ell(\underset{\sim}{x}',t')\right| \left|\frac{\partial f_o}{\partial v_\ell}\right| \quad . \tag{II.1.20}$$

Having established the criteria for linearization, either Eq. (II.1.4) or Eqs. (II.1.19) and (II.1.20), let us go back to Eq. (II.1.11). To go further, we would perform Laplace-in-time (L_p) and Fourier-in-space (F_r) transforms. In this respect, we can simply apply the transforms to Eq. (II.1.5). Thus, let

$$\delta f_1(\underset{\sim}{x},\underset{\sim}{v},t) = \int_{c_\omega} \frac{d\omega}{2\pi} \int_{-\infty}^{\infty} \frac{d^3\underset{\sim}{k}}{(2\pi)^3}\, \delta\hat{f}_1(\omega,\underset{\sim}{k}) \exp(-i\omega t + i\underset{\sim}{k}\cdot\underset{\sim}{x})$$

$$\equiv L_p^{-1}\, F_r^{-1}\, \delta\hat{f}_1(\omega,\underset{\sim}{k}) \quad . \tag{II.1.21}$$

Here, c_ω lies above the poles of $\delta f\ (\omega,\underset{\sim}{k})$ to ensure causality. Equation (II.1.5) then becomes

$$-i(\omega - \underset{\sim}{k}\cdot\underset{\sim}{v})\delta\hat{f}_1(\omega,\underset{\sim}{k}) = \delta\hat{f}_{10}(\underset{\sim}{k}) - \frac{q}{m}\,\delta\hat{\underset{\sim}{E}}(\omega,\underset{\sim}{k})\cdot\frac{\partial f_o}{\partial\underset{\sim}{v}} \quad . \tag{II.1.22}$$

Here, $\delta\hat{f}_{10}(\underset{\sim}{k}) = F_r\delta\hat{f}_1(\underset{\sim}{x},t = 0)$. In the electrostatic limit, we have

$$\delta\hat{\underset{\sim}{E}}(\omega,\underset{\sim}{k}) = -i\underset{\sim}{k}\ \delta\hat{\phi}(\omega,\underset{\sim}{k}) \quad . \tag{II.1.23}$$

We, thus, obtain

$$\delta\hat{f}_1(\omega,\underset{\sim}{k}) = \frac{i\delta\hat{f}_{10}-(q/m)\delta\hat{\phi}\ \underset{\sim}{k}\cdot(\partial f_o/\partial\underset{\sim}{v})}{\omega-\underset{\sim}{k}\cdot\underset{\sim}{v}} \quad . \tag{II.1.24}$$

Here, the dependence on ω, $\underset{\sim}{k}$ and/or $\underset{\sim}{v}$ has been surpressed. Noting that in Eq. (II.1.24), only the component of $\underset{\sim}{v}$ in the $\underset{\sim}{k}$ direction matters, we shall let

$$\underset{\sim}{k} = k\underset{\sim}{e}_k \quad , \tag{II.1.25}$$

$$\underset{\sim}{v} = u\underset{\sim}{e}_k + \underset{\sim}{v}_\perp \quad , \tag{II.1.26}$$

and

$$(\delta\hat{F}_1, F_o) \equiv \int d^2\underset{\sim}{v}_\perp \, (\delta\hat{f}_1, f_o) \quad . \tag{II.1.27}$$

Thus, from Eq. (II.1.25), we have

$$\delta\hat{F}_1 = \frac{i\delta\hat{F}_{10} - (q/m)\delta\hat{\phi}\; k(\partial F_o/\partial u)}{\omega - ku} \quad . \tag{II.1.28}$$

Equation (II.1.28) into the Poisson's equation

$$k^2\delta\hat{\phi} = 4\pi \sum_j N_{oj} q_j \int_{-\infty}^{\infty} du \; \delta\hat{F}_{1j} \quad , \tag{II.1.29}$$

we obtain

$$\delta\hat{\phi}(\omega,\underset{\sim}{k}) = \frac{4\pi i\left(\sum_j N_{oj}\; q_j \int du \; \delta\hat{F}_{10j}/(\omega - ku)\right)}{k^2 D(\omega,\underset{\sim}{k})} \quad , \tag{II.1.30}$$

where

$$D(\omega,\underset{\sim}{k}) = 1 + \sum_j \frac{\omega_{pj}^{\;2}}{k^2} \int_{-\infty}^{\infty} du \; \frac{k\partial F_{oj}/\partial u}{\omega - ku} \quad . \tag{II.1.31}$$

Thus,

$$\delta\phi(t,\underset{\sim}{k}) = L_p^{-1}\; \delta\hat{\phi}(\omega,\underset{\sim}{k}) = \int_{C_\omega} \frac{d\omega}{2\pi}\; \delta\hat{\phi}(\omega,\underset{\sim}{k})\exp(-i\omega t) \quad . \tag{II.1.32}$$

§II.2 Landau damping and phase mixing of ballistic modes

Combing Eqs. (II.1.30) and (II.1.32) , $\delta\phi(t,\underset{\sim}{k})$ can be written as

$$\delta\phi(t,\underset{\sim}{k}) = \frac{4\pi i}{k^2} \sum_j N_{oj}\, q_j \int_{-\infty}^{\infty} du\, \delta\hat{F}_{10j} \int_{C_\omega} \frac{d\omega}{2\pi} \exp(-i\omega t) \frac{1}{(\omega-ku)D(\omega,\underset{\sim}{k})} \qquad (II.2.1)$$

with D given by Eq. (II.1.31). For t>0, we close the C_ω contour at the lower half ω plane. We, thus, can pick up the pole contribution from ω=ku and D (ω, $\underset{\sim}{k}$) =0. The ω=ku pole gives a time response of the exp(-ikut) form and is called the ballistic mode. We shall discuss it in more details later. The $D(\omega,\underset{\sim}{k})$ = 0 yields the normal modes of the system. It is clear, however, since $D(\omega,\underset{\sim}{k})$ is analytic only for Im $\omega \geq$ Im c_ω, we need to analytically continuate it to the ω plane below c_ω. Referring to Eq. (II.1.31) and Fig. (II.2.1), it is clear that as Im $\omega \to 0^+$, $D(\omega,\underset{\sim}{k})$ as defined by Eq. (II.1.31) becomes non-analytic since the pole at ω/k approaches the integration contour which is the

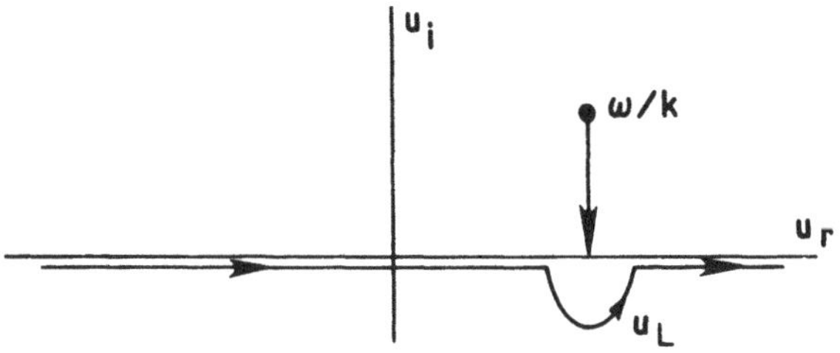

Fig. (II.2.1) Sketch of the deformed Landau contour in the u-plane integration.

real-u axis. Thus, the ω_r axis is the branch line of $D(\omega,\underset{\sim}{k})$. To analytically continuate $D(\omega,\underset{\sim}{k})$ for Im $\omega \leq 0$, we then need to deform the u-integration contour below the ω/k pole. This deformed contour, shown as u_L in Fig. (II.2.1), was first prescribed by Landau in his classical paper [J. Physics (U.S.S.R), 10, 25(1946)] where he resolved the singularity encountered in the u-intergration via the initial-value approach. The analytical continuation of D is thus given by

$$\hat{D}(\omega,\underset{\sim}{k}) = 1 + \sum_j \frac{\omega_{pj}^2}{k^2} \int_{u_L} du \, \frac{k\partial F_{oj}/\partial u}{\omega - ku} \quad . \tag{II.2.2}$$

An equivalent description is to replace $(u - \omega/k)^{-1}$ in Eq. (II.1.31) by $\overline{(u - \omega/k)^{-1}}$ such that

$$\overline{(u - \omega/k)^{-1}} = \begin{cases} (u - \omega/k)^{-1} & \text{for Im } (\omega/k) > 0, \\ P(u - \omega/k)^{-1} + i\pi\,\delta(u - \omega/k) & \text{for Im } (\omega/k) = 0 \quad , \\ P(u - \omega/k)^{-1} + 2\,i\pi\,\delta(u - \omega/k) & \text{for Im } (\omega/k) < 0 \quad ; \end{cases} \tag{II.2.3}$$

with P stands for the principal value.

We now examine qualitatively the ω poles given by $\hat{D}=0$. From Eq. (II.2.2) or (II.2.3), it is clear that

$$\hat{D} = \hat{D}_r + i\,\hat{D}_i \quad , \tag{II.2.4}$$

where $\hat{D}_i$ is due to the presence of the Landau pole. Since the Landau pole only involves the resonant particles with $u = \omega/k \equiv u_{ph}$, we can, in general, assume $|\hat{D}_i| << |\hat{D}_r|$. Correspondingly, we can assume $\omega = \omega_r + i\gamma$. From $\hat{D} = 0$, we have

$$\hat{D}_r(\omega_r,\underset{\sim}{k}) = 0 \quad , \tag{II.2.5}$$

which gives

$$\omega_r = \omega_{r\ell}\,(\underset{\sim}{k}) \quad , \tag{II.2.6}$$

and $\ell=1, \ldots, N$ is ℓth normal mode. In the next order, we find

$$\gamma_\ell = \frac{\hat{D}_i(\omega_\ell, \underset{\sim}{k})}{\partial\hat{D}_r/\partial\omega_{r\ell}} \quad . \tag{II.2.7}$$

Specifically, using Eq. (II.2.3), $\hat{D}_r$ and $\hat{D}_i$ are given by

$$\hat{D}_r(\omega_r, \underset{\sim}{k}) = 1 + \sum_j \frac{\omega_{pj}^2}{k^2} P\int_{-\infty}^{\infty} du \, \frac{k \, \partial F_{oj}/\partial u}{\omega_r - ku} \quad , \tag{II.2.8}$$

and

$$\hat{D}_i(\omega_r, \underset{\sim}{k}) = - i\pi \sum_j \frac{\omega_{pj}^2}{k^2} \frac{\partial F_{oj}}{\partial u}\bigg|_{u=\omega_r/k} \quad . \tag{II.2.9}$$

From Eq. (II.2.9), we see that the resonant wave-particle interaction can play the role of dissipation in collisionless plasmas. According to discussions in Sec. §I.2, it then can be expected that waves can either grow or decay depending on the signs of wave energies and collisionless dissipation; i.e. $\hat{D}_i$. This type of wave damping (growth) is called Landau damping (growth) for an obvious reason. It is interesting to note that the sign of $\hat{D}_i$ depends on the slope of velocity distribution function, i.e., $\partial F_o/\partial u$. This has a simple physical explanation because, in a plane wave, particles travelling faster (slowlier) than the wave phase velocity, ω_r/k, are decellerated (accellerated) by the wave and, hence, feed (extract) energy to (from) the wave. Refer to Ch. 7 of Stix's book for a physical picture of Landau damping.

Having found the zeroes of $\hat{D}$, we can go back to Eq. (II.II.1) and calculate $\delta\phi(t,\underset{\sim}{k})$. Since we are interested in time-asymptotic rather than transient responses, only the $\omega=ku$ pole and the normal mode with the maximum γ, denoted as ω_m, need to be kept. Hence,

$$\lim_{t\to\infty} \delta\phi(t,\underset{\sim}{k}) = \frac{4\pi}{k^2}\sum_j N_{oj}q_j\int_{u_L} du\ \delta\hat{F}_{1oj}\left[\frac{e^{-ikut}}{\hat{D}(ku,\underset{\sim}{k})} + e^{-i\omega_m t}\ \frac{1}{(\omega_m - ku)(\partial\hat{D}/\partial\omega_m)}\right] .$$

(II.2.10)

Now the first term in Eq. (II.2.10) corresponds to contributions from the ballistic modes. As $t \to \infty$, exp (-ikut) becomes highly oscillatory in u. Thus, given $\delta\hat{F}_{10j}$ sufficiently smooth in u, we expect the ballistic-mode contributions decay rapidly in time. This is called phase-mixing of the ballistic modes. Hence, if the system is marginally stable or unstable; i.e., Im $\omega_m \geq 0$, it is clear that the normal modes determine the time-asymptotic responses. If, however, the normal modes are damped (i.e., Im $\omega_m < 0$), the time-asymptotic responses will then depend on how fast the ballistic modes decay in time. In order to answer this question, let us consider the following two cases. First, we assume $\delta\hat{F}_{10}$ is an entire function;

$$\delta\hat{F}_{10} = A_o \exp\left(\frac{-(u-u_o)^2}{(\Delta u)^2}\right) \quad , \tag{II.2.11}$$

where A_o is a small but finite constant. To further simplify the analysis, we assume $|u_o| >> |\Delta u|,\ |\omega_r/k|$. We then have $\hat{D}(ku,k) \simeq 1$ and

$$\int_{u_L} du\ \delta\hat{F}_{10}\ e^{-ikut} \propto e^{-iku_o t}\ e^{-(k\Delta u)^2 t^2} \quad ; \tag{II.2.12}$$

and, hence, in this case the least damped normal mode determine the $t\to\infty$ response. Next, we consider a non-analytic $\delta\hat{F}_{10}$ given by

$$\delta\hat{F}_{10}(u) = A_o \begin{cases} 1 & , \text{ for } |u - u_o| \leq \Delta u \\ 0 & , \quad \text{otherwise.} \end{cases} \tag{II.2.13}$$

In this case, we have

$$\int_{u_L} du\ \delta\hat{F}_{10}\ e^{-ikut} \propto \frac{1}{t} \sin(k\Delta ut)\ e^{-iku_0 t} \quad ; \qquad (II.2.14)$$

and, thus, the ballistic modes, which only decay algebraically in t, dominate over the exponentially damped normal modes in the time-asymptotic responses. Now, in real physical situations, $\delta\hat{F}_{10}$ must be a smoothly varying function in u and has a sufficiently board Δu. It is in this perspective that we expect the normal modes to determine the time-asymptotic behaviours of the system. It is worthwhile to note that, while ballistic modes do not contribute to field perturbations, they do carry information about the initial phase-space perturbations. To see this point, note that, from Eq. (II.1.24) or, equivalently,

$$\left(\frac{\partial}{\partial t} + iku\right) \delta F_1(t,k) = \frac{q}{m}\ ik\delta\phi(t,k)\ \frac{\partial F_0}{\partial u} \quad ; \qquad (II.2.15)$$

we have

$$\delta F_1(t,k) = e^{-ikut} \left[\delta\hat{F}_{10} + i\,\frac{q}{m}\,k\,\frac{\partial F_0}{\partial u} \int_0^t \delta\phi(t',k)\ e^{ikut'}\ dt'\right] \quad . \qquad (II.2.16)$$

Equation (II.2.16) clearly indicates that initial phase-space perturbations, $\delta\hat{F}_{10}$, propagate without any damping even though the perturbed fields may decay away; i.e., $|\delta\phi| \to 0$ as $t \to \infty$. This conservation of information is, of course, due to the fact that the Vlasov plasma conserves entropy. We remark that δF_1 remains undamped is the basis for the interesting nonlinear phenomenon known as plasma echoes.

Finally, let me suggest two advanced readings on Landau damping. (1)

N.G. Van Kampen, Physica <u>21</u>, p.949 (1955). (2) K.M. Case Ann. Physics (N.Y.) <u>7</u>, p.349 (1959). Both papers require some mathematical inclination. The ballistic modes are also called Van Kampen modes. These modes can be attributed to the continuous spectrum. This concept of continuum is, surprisingly, often very useful in plasma physics research with regard to resonance heating, ideal magnetohydrodynamic waves in nonuniform plasmas, etc.

<u>Homework #4</u>

(1) Considering a cold plasma with two equal-density counter-streaming electron beams with velocity $\pm V_b$, respectively. In the electrostatic limit, show

(1.a) the linear dispersion relation is given by

$$1 - \frac{\omega_b{}^2}{(\omega-kV_b)^2} - \frac{\omega_b{}^2}{(\omega+kV_b^2)} = 0. \tag{H.4.1}$$

Here, $\omega_b{}^2 = 4\pi N_b e^2/m_e$ and $N_b = N_0/2$ with N_0 being the background ion density.

(1.b) The system is linearly unstable and plot the ω vs. k curves.

(1.c) The instability is absolute.

(1.d) By going to a moving reference frame via $x' = x + V_b t$ transform, show the instability becomes convective.

(2) Consider the weak beam-plasma instability with the dispersion relation

$$1 - \frac{\omega_p{}^2}{\omega^2} - \frac{\omega_b{}^2}{(\omega-kV_b)^2} = 0 \quad , \quad |\omega_b|^2 \ll |\omega_p|^2 \quad . \tag{H.4.2}$$

Given a steady-state driving source with $\omega = \omega_0$ at $x=0$ and letting $k = k_r + ik_i$, calculate

(2.a) The range of k_r that the source will be spatially amplified, and

(2.b) the corresponding spatial amplification factor; i.e., $|k_i\ x|$.

(3) Using Nyqiust technique to prove the Newcomb-Gardner theorem; i.e. the electron Langmuir (Bohm-Gross) wave is stable if F_{oe} (u) has only one peak (see the sketch).

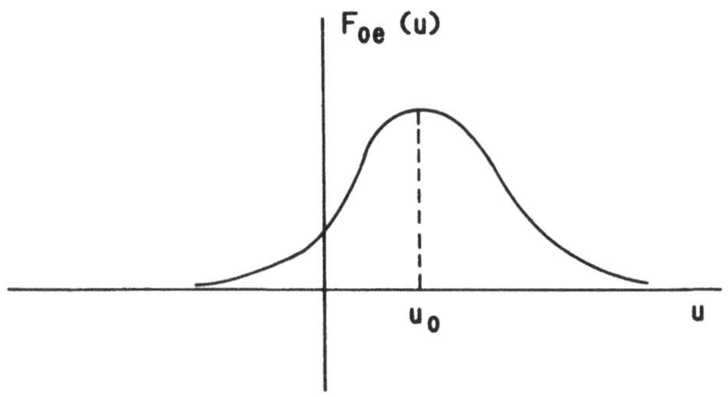

Fig. (H.4.1) Sketch of the electron distribution function, F_{oe}, for Problem (3) in Homework #4.

§II. 3 Particle trapping and the breakdown of linearization

So far we have considered linear wave dynamics via the linearized Vlasov equation. While the validity conditions for the linear theory are formally expressed in Eq. (II.1.4) or Eqs. (II.1.19) and (II.1.20), we would like to further quantify them for the simple case of a monochromatic travelling wave; i.e.,

$$\delta E(x,t) = \delta\hat{E}_o \exp(\gamma t)\sin(\omega_r t - kx) \quad . \tag{II.3.1}$$

Now from discussions of Sec. § II.1, we know that linearization is equivalent to approximate the exact particle orbits by the unperturbed orbits, i.e., linear theory assumes

$$|(\omega_k - ku)\delta\hat{F}_1| = \left|\frac{q}{m}\,\delta\hat{E}\,\frac{\partial F_o}{\partial u}\right| \gg \left|\frac{q}{m}\,\delta\hat{E}\,\frac{\partial \delta\hat{F}_1}{\partial u}\right| \quad . \tag{II.3.2}$$

Here, $\omega_k = \omega_r + i\gamma$ and $\delta\hat{E} = \delta\hat{E}_o \exp(\gamma t)$. It is, thus, clear that linearization will first breakdown for those particles with minimal $|\omega_k - ku|$; i.e., resonant particles with $u=u_r \equiv \omega_r/k$. Equation (II.3.2) then becomes

$$\left|\frac{\partial F_o}{\partial u}\right|_{u=u_r} \gg \left|\frac{q}{m}\,\delta\hat{E}\,\frac{\partial \delta\hat{F}_1}{\partial u}\right|_{u=u_r} \quad . \tag{II.3.3}$$

Noting $\delta\hat{F}_1 = -i(q/m)\,[\delta\hat{E}(\partial F_o/\partial u)]/(\omega_k - ku)$, Eq. (II.3.3) can be readily shown to yield

$$\gamma^2 \gg \left|\frac{qk\,\delta\hat{E}}{m}\right| \equiv \omega_b^2 \quad . \tag{II.3.4}$$

Here, in Eq. (II.3.4), ω_b corresponds to the bounce frequency of a charged particle deeply trapped in an electrostatic wave with amplitude $\delta\hat{E}$ and wave number k. Let us consider the trapping process in more details. Furthermore, we assume $\gamma=0$ and, hence, $\delta\hat{E}$ has a constant amplitude. The corresponding equation of motion of a charged particle then is

$$\frac{d^2x}{dt^2} = -\left(\frac{q}{m}\right)\delta\hat{E}_o \sin(kx - \omega_r t) \quad . \tag{II.3.5}$$

Changing to a frame moving with the phase velocity of the wave; i.e., letting

$$\delta x = x - (\omega_r/k)t \quad , \tag{II.3.6}$$

Eq. (II.3.5) becomes

$$\frac{d^2\delta x}{dt^2} = -\left(\frac{q}{m}\right)\,\delta\hat{E}_o\,\sin(k\delta x) \quad . \tag{II.3.7}$$

Equation (II.3.7) can be exactly solved in term of the elliptic integrals. For the present purpose, we consider those deeply trapped particles; i.e., $|k\delta x| << 1$. Equation (II.3.7) then reduces to

$$\frac{d^2\delta x}{dt^2} = -\omega_{bo}^2\,\delta x \quad . \tag{II.3.8}$$

Equation (II.3.8) clearly exhibits the bounce motion of particles trapped in a travelling electrostatic wave. The phase-space ($\delta v = \delta\dot{x}$, δx) plot for Eq. (II.3.7) is also sketched in Fig. (II.3.1).

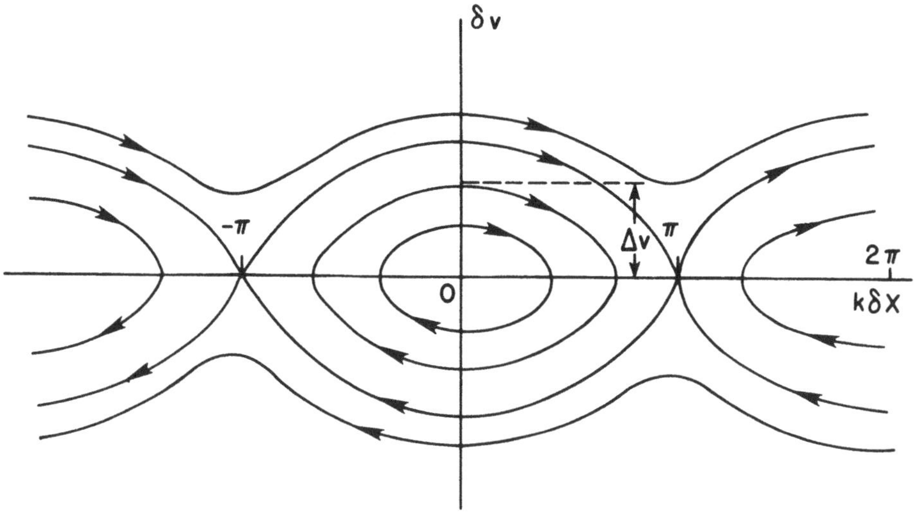

Fig. (II.3.1) Sketch of the ($\delta v,\delta x$) phase-space plot for Eq. (II.3.7).

In Fig. (II.3.1.), $\delta v = v - \omega_r/k$, and Δv is sometimes called the trapping width. From Eq. (II.3.7), it is easy to see that

$$\Delta v = \left[\frac{2q\delta\hat{E}_o k}{mk^2} \right]^{1/2} = \sqrt{2} \; \frac{\omega_{bo}}{k} \quad . \tag{II.3.9}$$

From the particle trapping picture, it is then straightforward to develop a physical explanation for the validity condition givey by Eq. (II.3.4). Noting that Landau damping is due to the presence of more faster, with respect to the wave, resonant particles than slowlier ones. On the other hand, the resonant particles change their relative velocities with respect to the wave on the characteristic time scale of ω_b^{-1}. Thus, the time-asymptotic linear theory is meaningful only if the wave can decay several Landau-damping time, γ^{-1}, before particle can execute a bounce motion. Similar conclusions can, of course, be drawn for the case with Landau growth. In the case of damping, it is clear that, if Eq. (II.3.4) holds initially, it will be valid for all time. In the case of instability, since $\delta\hat{E}$ and, hence, ω_b increase with time, we know that, sooner or later, Eq. (II.3.4) will be violated and, thereby, nonlinear treatment is called for.

Let us now consider the opposite limit; i.e., $|\gamma| \ll \omega_{bo}$. In this limit, time-asymptotic linear theory becomes meaningless. So we consider the initial-value problem. Furthermore, we let $t \ll 1/|\gamma|$ so that δE in Eq. (II.3.1) has a constant amplitude. From the Poisson's equation, we know that $\delta E(x,t=0)$ is consistent with an initial perturbed velocity distribution function of the form

$$\delta F(x,u,t=0) = \delta F_o'(u) \cos kx \quad . \tag{II.3.10}$$

From Eq. (II.1.14), we then find

$$\delta F(x,u,t) = \delta F_0(u) \cos (kx - kut) + \left(\frac{q}{m}\right) \int_0^t dt' \, \delta\hat{E}_0 \sin (kx' - \omega_r t') \frac{\partial F_0}{\partial u} . \tag{II.3.11}$$

Here, $x' = x - u(t-t')$. Equation (II.3.11) can be readily integrated to yield

$$\delta F(x,u,t) = \delta F_0(u) \cos(kx-kut) + \frac{q}{m} \delta\hat{E}_0 \frac{\partial F_0}{\partial u} \frac{\cos (kx - \omega_r t) - \cos (kx - kut)}{\omega_r - ku} . \tag{II.3.12}$$

Equation (II.3.12) clearly indicates a possible singularity at $u=\omega_r/k$; i.e., the resonant particles. Expanding about $u=u_r \equiv \omega_r/k$ and taking the velocity differentiation, we find that $\partial\delta F/\partial u|_{u=u_r}$ becomes

$$\frac{\partial \delta F}{\partial u}\bigg|_{u=u_r} \approx \delta F_0(u_r) kt \sin(kx-ku_r t) - \frac{1}{2} \frac{\partial F_0}{\partial u} (t\omega_{bo})^2 \cos(kx - \omega_r t) . \tag{II.3.13}$$

Equation (II.3.13), thus, becomes secular in time and the linear theory breaks down for $t \gtrsim \omega_{bo}$. For more detailed analysis of particle-trapping effects on Landau damping, we refer to the paper by T. M. O'neil, Phys. Fluids, 8, 2255(1965).

§II.4 Normal modes

Our discussions here are the standard ones and, hence, will be brief. The readers are refered to typical text books such as Krall and Trivelpiece for more detailed analyses. These normal modes, of course, correspond to $\hat{D} = 0$ with $\hat{D}$ given by Eq. (II.2.2) or, equivalently Eqs. (II.2.8) and (II.2.9).

First, in the high-frequency electrostatic regime, we have the Bohm-Gross (warm-electron Langmuir) mode. Here, $|\omega| \sim \omega_{pe} \gg \omega_{pi}, kv_{te}, kv_{ti}$. More precisely, we have

$$\omega^2 \simeq \omega_{pe}{}^2(1 + 3k^2\lambda_{De}^2) \quad , \tag{II.4.1}$$

with $\lambda_{De}{}^2 = v_{te}{}^2/2\omega_{pe}{}^2$ and $v_{te}{}^2/2 \equiv \int_{-\infty}^{\infty} u^2F_{oe}\, du$. For a monotonically decreasing F_{oe}, such as a Maxwellian distribution $F_{oe}=(1/\sqrt{\pi}v_{te})\exp(-u^2/v_{te}{}^2)$, there is weak Landau damping.

Next, there is the low-frequency electrostatic ion-acoustic wave, which has no counterpart in the cold plasma description. This wave has an intermediate phase velocity; i.e. $v_{te} \gg \omega_r/k \gg v_{ti}$. We further take $F_{oe}=F_{oe}[(u/v_{te})^2]$. We, thus, find from Eq. (II.2.8)

$$\hat{D}_r \simeq 1 - \frac{\omega_{pi}^2}{\omega^2} + \frac{1}{k^2\lambda_{De}^2} = 0 \quad , \tag{II.4.2}$$

or

$$\omega^2 = \frac{k^2C_s^2}{1+k^2\lambda_{De}^2} \quad . \tag{II.4.3}$$

Here, $C_s{}^2=\omega_{pi}{}^2\lambda_{De}{}^2=T_e/M_i$. We note that, to satisfy the $\omega_r/k \gg v_{ti}$ condition, we require $T_e \gg T_i$. For $T_e \sim T_i$, the ion-acoustic wave is heavily ion Landau damped. For $T_e \gg T_i$, it suffers mainly weak electron Landau damping.

At this point, let us introduce the terminology of susceptibility, χ_j, for the jth species, which is defined as

$$\chi_j\, \delta\hat{\phi} = -4\pi \frac{\delta\hat{\rho}_j}{k^2} \quad . \tag{II.4.4}$$

Thus, the electrostatic dielectric constant becomes

$$\hat{D} = 1 + \sum_j \chi_j \quad . \tag{II.4.5}$$

For ion-acoustic waves and, generally, for low-frequency waves, we have $|\chi_j| >> 1$ and, hence, the dispersion relation $\hat{D} = 0$ becomes, approximately, $\sum_j \chi_j = 0$ or

$$\sum_j \delta\hat{\rho}_j = 0 \quad . \tag{II.4.6}$$

Equation (II.4.6) is often called the quasi-neutrality condition. For ion-acoustic waves, its validity requires $k^2\lambda_{De}^2 << 1$.

Now it is interesting to calculate the wave energies for both the Bohm-Gross and ion-acoustic waves in the $k^2\lambda_{De}^2 << 1$ limit. For the Bohm-Gross (electron Langmuir) oscillations, we have

$$\delta W_{BG} \simeq \frac{|\delta\hat{E}|^2}{16\pi} (1 + 1) \quad . \tag{II.4.7}$$

For the ion-acoustic wave, we find

$$\delta W_{IA} \simeq \frac{|\delta\hat{E}|^2}{8\pi} (1 + 1/k^2\lambda_{De}^2) \simeq \frac{|\delta\hat{E}|^2}{8\pi} \frac{1}{k^2\lambda_{De}^2} \quad . \tag{II.4.8}$$

Thus, in the case of ion-acoustic waves, we have particle kinetic energy dominates over the electric field energy and this is, of course, related to the fact that the waves have such slow phase velocities that the quasi-neutrality condition is maintained.

At this juncture, let us briefly introduce the plasma dispersion function or, sometimes also called, the Z function defined as

$$Z(\xi) = \pi^{-1/2} \int_{u_L} \frac{dx \exp(-x^2)}{x-\xi} \quad . \tag{II.4.9}$$

This function is useful for Maxwellian distribution functions. It has been tabulated by Fried and Conte ["The Plasma Dispersion Function", Academic Press, N.Y. (1961)] and there exists standard computer subroutine. Thus, for $F_0 = (1/\sqrt{\pi}V_t)\exp(-u^2/V_t^2)$, the electrostatic susceptibility as defined by Eqs. (II.4.5) and (II.2.2) becomes

$$\chi_j = (k\lambda_{Dj})^{-2}[1 + \xi_j Z(\xi_j)] \quad , \tag{II.4.10}$$

where $\xi_j = \omega/kV_{tj}$.

Finally, for isotropic distributions, i.e., $f_0=f_0(v^2)$, we have the electromagnetic normal mode with the following dispersion relation

$$\omega^2 = \omega_{pe}^2 + c^2k^2 \quad . \tag{II.4.11}$$

Note, since $\omega/k > c$ (speed of light), this mode has no collisionless wave-particle resonance. It can, thus, be damped only by collisional dissipations.

§II.5 Instabilities

(i) Warm beam-plasma Instability. For simplicity, we shall assume cold background electrons but a warm beam. The corresponding electrostatic dispersion relation is then

$$1 - \frac{\omega_{pe}^2}{\omega^2} - \frac{\omega_{pb}^2}{k^2} \int_{u_L} du \frac{k\ \partial F_B/\partial u}{ku-\omega} = 0 \quad . \tag{II.5.1}$$

Here, ω_{pb} is the beam plasma frequency. To further make Eq. (II.5.1) analytically tractable, we shall take F_B to be Lorentzian [O'Neil and Malmberg, Phys. Fluids, 11, 1754 (1968)]; i.e.,

$$F_B = (\frac{u_{tb}}{\pi})[(u - u_b)^2 + u_{tb}{}^2]^{-1} \quad . \tag{II.5.2}$$

Equation (II.5.1) then becomes

$$1 - \frac{\omega_{pe}{}^2}{\omega^2} - \frac{\omega_{pb}{}^2}{(\omega - ku_b + iku_{tb})^2} = 0 \quad . \tag{II.5.3}$$

Equation (II.5.3) can be readily analyzed in the various limits of the parameter $|\omega_{pe}/ku_b|$.

First, for $|ku_b| \gg \omega_{pe}$, we find, for the electron plasma wave,

$$\omega_1 = \omega_{pe} + \delta\omega_1 \quad , \tag{II.5.4}$$

and

$$\frac{2\delta\omega_1}{\omega_{pe}} \simeq (\frac{\omega_{pe}}{ku_b})^2(1 + \frac{2i\, u_{tb}}{u_b}) \quad . \tag{II.5.5}$$

Thus, in contrast to the cold beam results, the electron plasma wave is unstable in this case. This instability can be understood in terms of destabilization of a positive-energy wave by negative collisionless dissipation. Meanwhile, for the two beam modes $(|\omega| \sim |ku_b|)$, we find

$$\omega \simeq ku_b \pm \omega_{pb} - ik\, u_{tb} \quad ; \tag{II.5.6}$$

i.e. they are stable. Now the stability may be regarded as the effect of negative (positive) dissipation on negative (positive) - energy waves.

In the opposite limit, $|ku_b| << \omega_{pe}$, we have, for the electron plasma wave, Eq. (II.5.4) but with $\delta\omega_1$ given by

$$\frac{2\delta\omega_1}{\omega_{pe}} \approx \left(\frac{\omega_{pb}}{\omega_{pe}}\right)^2\left(1 - \frac{2iku_{tb}}{\omega_{pe}}\right) \quad . \tag{II.5.7}$$

Thus, it is stable due to the effect of positive dissipation on a positive-energy wave. As to the beam modes, we find

$$\omega \approx ku_b - iku_{tb} \pm \frac{i(\omega_{pb}ku_b)}{\omega_{pe}} \quad . \tag{II.5.8}$$

Hence, for

$$u_{tb} > u_b \frac{\omega_{pb}}{\omega_{pe}} = u_b\left(\frac{N_b}{N_o}\right)^{1/2} \quad , \tag{II.5.9}$$

all the modes are stable in this long wavelength limit.

Finally, we consider the strongly-coupled regime; i.e., $k_o u_b = \omega_{pe}$. Letting $\omega = \omega_{pe} + \delta\omega$, Eq. (II.5.3) then yields

$$\frac{2\delta\omega}{\omega_{pe}} = \frac{\omega_{pb}^2}{(\delta\omega + ik_o u_{tb})^2} \quad . \tag{II.5.10}$$

Equation (II.5.10) shows stability for

$$u_{tb} > u_b\left(\frac{N_b}{2N_o}\right)^{1/3} \equiv u_c \quad . \tag{II.5.11}$$

Noting that Eq. (II.5.11) automatically gurantees Eq. (II.5.9), it is then obvious that u_c, as given by Eq. (II.5.11), is the characteristic beam thermal spread signifying the transition from the cold beam ($u_{tb} < u_c$) case to the warm beam ($u_{tb} > u_c$) case. The dispersion curves for the two cases are sketched in Fig. (II.5.1).

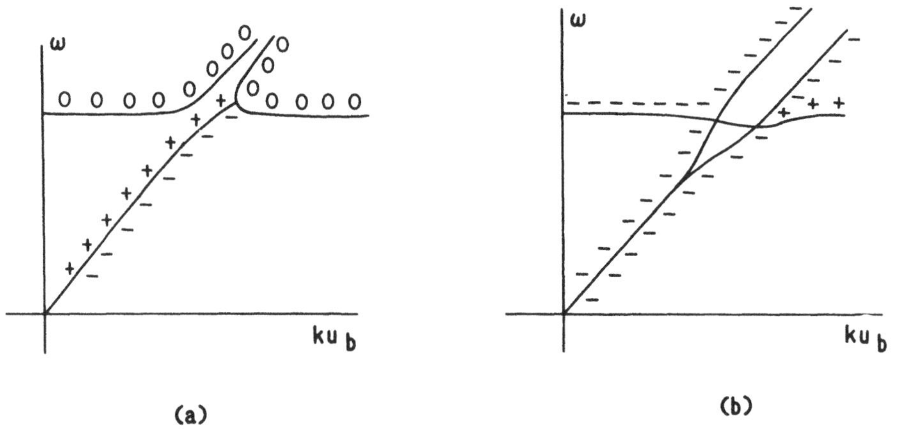

Fig. (II.5.1) Sketches of the ω vs. ku_b curves for the Lorentzian beam-plasma dispersion relation of Eq. (II.5.3) in (a) the cold beam limit and (b) the warm beam limit. Here, +,0, and - denote, respectively, Imω > 0, Imω = 0, and Imω < 0.

(ii) Electromagnetic instability due to velocity-space anisotropy (Weibel instability). The polarization of the wave considered here is sketched in Fig. (II.5.2).

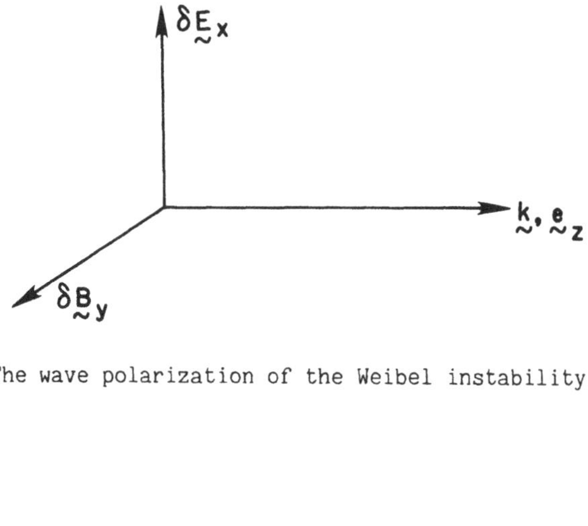

Fig. (II.5.2) The wave polarization of the Weibel instability.

The linearized Vlasov equation

$$i(\omega - kV_z)\delta\hat{f}_i = \frac{q}{m}\left(\delta\hat{\underset{\sim}{E}} + \frac{\underset{\sim}{V}\times\delta\hat{\underset{\sim}{B}}}{c}\right)\cdot\frac{\partial f_o}{\partial \underset{\sim}{V}} \quad \text{(II.5.12)}$$

then yields

$$\delta\hat{f}_1 = -i\,\frac{q}{m}\,\frac{\delta\hat{E}_x}{(\omega - kV_z)}\left[\left(1 - \frac{kV_z}{\omega}\right)\frac{\partial}{\partial V_x} + \frac{kV_z}{\omega}\frac{\partial}{\partial V_z}\right]f_o. \quad \text{(II.5.13)}$$

With $f_o = f_o(V_x{}^2, V_y{}^2, V_z{}^2)$, it is clear from Eq. (II.5.13) that only δJ_x contributes and the dispersion relation is given by

$$-c^2k^2 + \omega^2 D_{xx} = 0 \quad , \quad \text{(II.5.14)}$$

where

$$D_{xx} = 1 + \frac{i4\pi\sigma_{xx}}{\omega} = 1 + \sum_j \frac{\omega_{pj}{}^2}{\omega^2}\left(\int d^3\underset{\sim}{V}\,\frac{V_x{}^2 k\partial f_o/\partial V_z}{\omega - kV_z} - 1\right) \quad . \quad \text{(II.5.15)}$$

If we further assume $|\omega| \gg |kV_z|$ and define

$$\int d^3\underset{\sim}{V}\, V_x{}^2\, f_o \equiv \frac{V_{tx}^2}{2} \quad , \quad \text{(II.5.16)}$$

Eq. (II.5.14) then becomes (with $m_i \to \infty$)

$$-c^2k^2 + \omega^2 - \frac{\omega_{pe}{}^2}{\omega^2}\frac{k^2V_{tx}^2}{2} - \omega_{pe}^2 = 0 \quad . \quad \text{(II.5.17)}$$

Equation (II.5.17) is quadratic in ω^2 and one readily obtains

$$\omega^2 = \{c^2k^2 + \omega_{pe}^2 \pm [(c^2k^2 + \omega_{pe}^2) + 2\omega_{pe}^2 k^2 V_{tx}^2]^{1/2}\}/2 \quad . \tag{II.5.18}$$

The minus sign in Eq. (II.5.18) gives $\omega^2<0$ and, hence, a purely growing instability. The instability growth rate, γ, is sketched in Fig. (II.5.3).

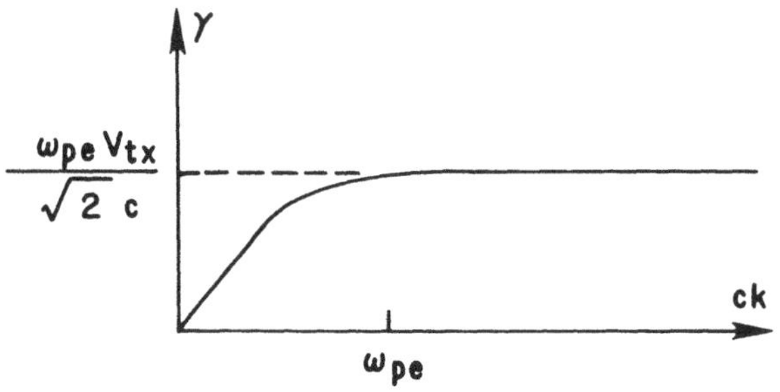

Fig. (II.5.3) Sketch of the growth rate (γ) of the Weibel instability vs. ck.

We conclude this disucssion by developing a physical picture for the Weibel instability. For this purpose, we may take the electrons to be current sheets with the following equilibrium distribution

$$f_o(\underset{\sim}{v}) = (\tfrac{1}{2})\delta(V_y)\,\delta(V_z)[\delta(V_x - V_o) + \delta(V_x + V_o)] \quad . \tag{II.5.19}$$

Now, at t = 0, let us give the system an initial perturbation of the form

$$\delta B_y(z, t = 0) = \delta B_o \sin kz \quad . \tag{II.5.20}$$

In response to δB_y, the electrons will move according to the equation of motion

$$\frac{dV_z}{dt} = \frac{qV_x}{mc}\,\delta B_y \quad . \tag{II.5.21}$$

In a time step Δt, we then have

$$(\Delta V_z)_\pm = \pm \left(\frac{qV_o}{mc}\right)\Delta t \ \delta B_y \quad . \qquad \text{(II.5.22)}$$

Here, the ± sign corresponds to $V_x = \pm V_o$[c.f. Fig. (II.5.4)].

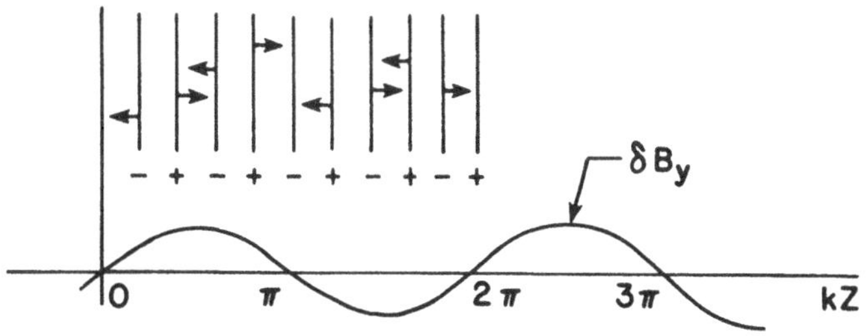

Fig. (II.5.4) Physical model of the Weibel instability in terms of $V_x = \pm V_o$ current sheets.

From Fig. (II.5.4), it is clear that the current sheets will start pinching; i.e., there will be more positive current sheets at certain regions in space (e.g., about $kz = \pi$) and more negative current sheets at other regions in space (e.g., about $kz = 0$). We, therefore, expect perturbations in J_x and B_y defined, respectively, as ΔJ_x and ΔB_y. If ΔB_y, is in phase with δB_y, the initial perturbation, we then have a positive feedback; i.e., an instability.

To see it more specifically, we use the continuity equation to calculate the density perturbations in the current sheets. We find

$$(\Delta N)_\pm = - \Delta t \frac{N_o}{2} \frac{\partial}{\partial z} (\Delta V_z)_\pm$$

$$= \mp(\Delta t)^2 \frac{N_o}{2} \frac{qV_o k}{mc} \delta B_o \cos kz \quad . \qquad \text{(II.5.23)}$$

The current density perturbation, ΔJ_x, is then given by

$$\Delta J_x = qV_o[\Delta N_+ - \Delta N_-]$$

$$= -\frac{N_o q^2 V_o^2 k}{mc} (\Delta t)^2 \; \delta B_o \cos kz \quad . \tag{II.5.24}$$

From the Ampere's law

$$\underset{\sim}{\nabla} \times \Delta \underset{\sim}{B} = \frac{4\pi}{c} \Delta \underset{\sim}{J} \tag{II.5.25}$$

it is easy to show that

$$\Delta B_y = \left(\frac{\omega_{pe} \, \Delta t \, V_o}{c}\right)^2 \delta B_y \quad . \tag{II.5.26}$$

Thus, ΔB_y reinforces δB_y and the initial perturbation further grows.

Homework #5

(1) In class, we show, using a Lorentzian beam distribution function, $F_B = (u_{tb}/\pi)[(u - u_b)^2 + u_{tb}^2]^{-1}$, that for $u_{tb} > u_b(N_B/N_o)^{1/3}$ the instability makes a transition from the cold beam-plasma instability to the warm beam-plasms (Bump-in-tail) instability. Show this conclusion also holds for a beam with a drifting Maxwellian distribution

$$F_B = \left(\frac{1}{\sqrt{\pi}\, u_{tb}}\right)\exp\left[-\left(\frac{u-u_b}{u_{tb}}\right)^2\right] \quad . \tag{H.5.1}$$

(2) Show that in an unmagnetized plasma, the dispersion tensor, $\underset{\approx}{\varepsilon}$, with $k = k\underset{\sim}{e}_z$ is given by

$$\underset{\approx}{\varepsilon} = \begin{vmatrix} \varepsilon_{xx} & 0 & 0 \\ 0 & \varepsilon_{yy} & 0 \\ 0 & 0 & \varepsilon_{zz} \end{vmatrix} , \tag{H.5.2}$$

where

$$\varepsilon_{\substack{x\\y}\substack{x\\y}} = -k^2 + \frac{\omega^2}{c^2}\left[1 - \sum_j \frac{\omega_{pj}^2}{\omega^2} + \sum_j \frac{\omega_{pj}^2}{\omega^2}\int d^3\underset{\sim}{v}\, \frac{k\partial f_o/\partial V_z}{\omega - kV_z}\, V_{\substack{x\\y}}^2\right] , \tag{H.5.3}$$

$$\varepsilon_{zz} = \frac{\omega^2}{c^2} D_{zz} = \frac{\omega^2}{c^2}\left(1 - \sum_j \frac{\omega_{pj}^2}{k^2}\int d^3V\, \frac{\partial f_o/\partial V_z}{V_z - \omega/k}\right) . \tag{H.5.4}$$

(3) Consider the Weibel instability ($\underset{\sim}{k} = k\, \underset{\sim}{e}_z$).

(3.a) By keeping the $O(k^2 V_{tz}{}^2/\omega^2)$ correction, show that finite V_{tz} is stabilizing.

(3.b) Demonstrate the V_{tz} stabilizing effect by examining this instability in the $|\omega| \ll |kV_{tz}|$ limit. For this purpose, assuming

$$f_{oe}(\underset{\sim}{v}) = f_{o\perp}(V_x, V_y)\, f_{oz}(V_z) , \tag{H.5.5}$$

with

$$f_{oz}(V_z) = \frac{1}{\sqrt{\pi}\, V_{tz}} \exp\left(-\frac{V_z^2}{V_{tz}^2}\right). \tag{H.5.6}$$

(3.c) Given f_{oe} as a bi-Maxwellian; i.e. referring to Eqs. (H.5.5), (H,5.6) and

$$f_{o\perp}(V_x, V_y) = \frac{1}{\pi V_{t\perp}^2} \exp\left(-\frac{V_x^2 + V_y^2}{V_{t\perp}^2}\right) , \tag{H.5.7}$$

obtain, via the Nyquist technique, the critical $\left(V_{t\perp}^2/V_{tz}^2\right)$ value for the instability.

(4) Assuming electron distribution is a drifting Maxwellian

$$F_{oe} = \frac{1}{\sqrt{\pi}}\frac{1}{U_{et}}\exp\left(-\frac{(U-U_o)^2}{U_{et}^2}\right) \quad , \qquad \text{(H.5.8)}$$

(4.a) for cold ions $T_i = 0$, what is the critical U_o for the ion-acoustic wave to be unstable (current-driven ion-acoustic instability)?

(4.b) For $T_e = T_i$, is there any electrostatic instability? Substantiate your answer.

CHAPTER III

Linear Waves and Instabilities in Uniform Magnetized Plasmas

§III.1. Magnetized Vlasov equation with the guiding-center transformation

In a magnetized plasma, the corresponding Vlasov equation is given by

$$\pounds_o f \equiv \left(\frac{\partial}{\partial t} + \underset{\sim}{v} \cdot \frac{\partial}{\partial \underset{\sim}{x}} + \frac{q}{mc} \underset{\sim}{v} \times \underset{\sim}{B}_o \cdot \frac{\partial}{\partial \underset{\sim}{v}}\right) f = -\frac{q}{m}\left(\delta \underset{\sim}{E} + \frac{\underset{\sim}{v}\times\delta\underset{\sim}{B}}{c}\right) \cdot \frac{\partial f}{\partial \underset{\sim}{v}} \quad . \tag{III.1.1}$$

Here, we assume there is no equilibrium electric field $\underset{\sim}{E}_o=0$. Generalization to $\underset{\sim}{E}_o \neq 0$ is not difficult. Now letting

$$f = f_0 + \delta f \quad , \tag{III.1.2}$$

we then have, for f_o,

$$\pounds_o f_0 = 0 \quad , \tag{III.1.3}$$

and, for the linear perturbed distribution function δf,

$$\pounds_o \delta f = -\frac{q}{m}\left(\delta \underset{\sim}{E} + \frac{\underset{\sim}{v}\times\delta\underset{\sim}{B}}{c}\right) \cdot \frac{\partial f_0}{\partial \underset{\sim}{v}} \quad ; \tag{III.1.4}$$

or, formally,

$$\delta f = \pounds_o^{-1}\left(-\frac{q}{m}\right)\left(\delta \underset{\sim}{E} + \frac{\underset{\sim}{v}\times\delta\underset{\sim}{B}}{c}\right) \cdot \frac{\partial f_0}{\partial \underset{\sim}{v}} \quad . \tag{III.1.5}$$

In Eq. (III.1.5), $\pounds_o^{-1}$ should be interpreted as integration along the particle's unperturbed orbit in the $(\underset{\sim}{x}, \underset{\sim}{v})$ phase space. Meanwhile, f_0, as given by Eq. (III.1.3), must be a function of particle's unperturbed constants of motion. Thus, the solutions of both f_0 and δf require detailed knowledge of the particle's unperturbed motion. It, furthermore, indicates that, if we can transform to coordinates where the unperturbed motion is simpler, the procedure involved in solving f_0 and δf will also be greatly simplified.

Based on the above intuitions and knowing that, in magnetized plasmas particle dynamics is simpler with the guiding-center picture, it is, therefore, suggestive to make a transformation from the particle phase space, $(\underset{\sim}{x}, \underset{\sim}{v})$, to the guiding-center phase space, $(\underset{\sim}{X}, \underset{\sim}{V})$, defined by

$$\underset{\sim}{X} = \frac{\underset{\sim}{x}+\underset{\sim}{v}\times\underset{\sim}{e}_{\parallel}}{\Omega} \quad , \qquad \text{(III.1.6)}$$

$$\underset{\sim}{V} = (\varepsilon,\mu,\alpha,\sigma) \quad , \qquad \text{(III.1.7)}$$

where $\underset{\sim}{e}_{\parallel}=\underset{\sim}{B}_0/B_0$, $\Omega = qB_0/mc$, $\varepsilon = v^2/2$, $\mu = v_{\perp}^{2}/2B_0$, $\sigma = \mathrm{sgn}(v_{\parallel})$, α is the gyrophase angle. Thus,

$$v_{\parallel} = \sigma\left(2(\varepsilon - \mu B_0)\right)^{1/2} \quad , \qquad \text{(III.1.8)}$$

$$\underset{\sim}{v}_{\perp} = v_{\perp}(\cos\alpha\ \underset{\sim}{e}_1 + \sin\alpha\ \underset{\sim}{e}_2) \quad , \qquad \text{(III.1.9)}$$

and $\underset{\sim}{e}_1$ and $\underset{\sim}{e}_2$ are two orthogonal unit vectors perpendicular to $\underset{\sim}{B}_0$; $\underset{\sim}{e}_1 \cdot \underset{\sim}{e}_{\parallel} = 0$ and $\underset{\sim}{e}_2 = \underset{\sim}{e}_{\parallel} \times \underset{\sim}{e}_1$.

From Eqs. (III.1.6) and (III.1.7), we have

$$\frac{\partial}{\partial \underset{\sim}{x}} \to \frac{\partial}{\partial \underset{\sim}{X}} \quad , \qquad \text{(III.1.10)}$$

$$\frac{\partial}{\partial \underset{\sim}{v}} \rightarrow \frac{\partial}{\partial \underset{\sim}{V}} + \frac{\underset{\approx}{I} \times \underset{\sim}{e}_{\parallel}}{\Omega} \cdot \frac{\partial}{\partial \underset{\sim}{X}} \quad , \tag{III.1.11}$$

$$\frac{\partial}{\partial \underset{\sim}{V}} \rightarrow \underset{\sim}{v} \frac{\partial}{\partial \varepsilon} + \underset{\sim}{v}_{\perp} \frac{\partial}{B_o \partial \mu} + \underset{\sim}{e}_{\alpha} \frac{1}{v_{\perp}} \frac{\partial}{\partial \alpha} \quad , \tag{III.1.12}$$

and

$$\int d^3 \underset{\sim}{v} \rightarrow \sum_{\sigma} \int \frac{B_o d\mu d\varepsilon d\alpha}{|v_{\parallel}|} \quad . \tag{III.1.13}$$

Here $\underset{\sim}{e}_{\alpha} = \underset{\sim}{e}_{\parallel} \times \underset{\sim}{v}_{\perp}/v_{\perp}$. The unperturbed Vlasov propagator, $\pounds_o$, then becomes, term by term

$$\frac{\partial}{\partial \underset{\sim}{t}} \rightarrow \frac{\partial}{\partial t} \quad , \tag{III.1.14}$$

$$\underset{\sim}{v} \cdot \frac{\partial}{\partial \underset{\sim}{x}} \rightarrow v_{\parallel} \frac{\partial}{\partial X_{\parallel}} + \underset{\sim}{v}_{\perp} \cdot \frac{\partial}{\partial \underset{\sim}{X}_{\perp}} \quad , \tag{III.1.15}$$

$$\frac{q}{mc} \underset{\sim}{v} \times \underset{\sim}{B}_o \cdot \frac{\partial}{\partial \underset{\sim}{v}} \rightarrow \underset{\sim}{v} \times \underset{\sim}{\Omega} \cdot \left(\underset{\sim}{e}_{\alpha} \frac{1}{v_{\perp}} \frac{\partial}{\partial \alpha} + \underset{\approx}{I} \times \frac{\underset{\sim}{e}_{\parallel}}{\Omega} \cdot \frac{\partial}{\partial \underset{\sim}{X}} \right)$$

$$= - \Omega \frac{\partial}{\partial \alpha} - \underset{\sim}{v}_{\perp} \cdot \frac{\partial}{\partial \underset{\sim}{X}_{\perp}} \quad . \tag{III.1.16}$$

That is,

$$\pounds_o \rightarrow \pounds_g \equiv \frac{\partial}{\partial t} + v_{\parallel} \frac{\partial}{\partial X_{\parallel}} - \Omega \frac{\partial}{\partial \alpha} \quad . \tag{III.1.17}$$

The equilibrium equation, Eq. (III.1.3), becomes

$$\pounds_g f_{og} = - \Omega \frac{\partial}{\partial \alpha} f_{og} = 0 \quad . \tag{III.1.18}$$

Here, subscript g denotes a quantity of the ($\underset{\sim}{X}$, $\underset{\sim}{V}$) variables. Hence, for a uniform magnetized plasma,

$$f_{og} = f_{og}(\varepsilon,\mu,\sigma) \qquad \text{(III.1.19)}$$

or

$$f_o = f_o(v_\perp, v_\parallel) \quad . \qquad \text{(III.1.20)}$$

§III.2 Solution of the linearized equation in the electrostatic limit

In the electrostatic limit, the linearized Vlasov equation, Eq. (III.1.4), transforms to

$$\pounds_g \delta f_g = \left(\frac{q}{m}\right)\left(\frac{\partial \delta\phi_g}{\partial \underset{\sim}{X}}\right)\cdot\left[v_\parallel \underset{\sim}{e}_\parallel \frac{\partial}{\partial \varepsilon} + \underset{\sim}{v}_\perp\left(\frac{\partial}{\partial \varepsilon} + \frac{\partial}{B_o \partial \mu}\right)\right]f_{og} \quad . \qquad \text{(III.2.1)}$$

Noting that for any field variable, e.g. $\delta\phi$, we have

$$\frac{\partial \delta\phi}{\partial \underset{\sim}{V}} = 0 \quad ; \qquad \text{(III.2.2)}$$

i.e.,

$$\left(\underset{\sim}{v}\frac{\partial}{\partial \varepsilon} + \underset{\sim}{v}_\perp \frac{\partial}{B_o \partial \mu} + \underset{\sim}{e}_\alpha \frac{1}{v_\perp}\frac{\partial}{\partial \underset{\sim}{\alpha}} + \underset{\approx}{I} \times \underset{\sim}{e}_\parallel \cdot \frac{1}{\Omega}\frac{\partial}{\partial \underset{\sim}{X}}\right)\delta\phi_g = 0 \quad . \qquad \text{(III.2.3)}$$

From the $\underset{\sim}{e}_\alpha$ component of Eq. (III.2.3), we obtain

$$\underset{\sim}{v}_\perp \cdot \frac{\partial \delta\phi_g}{\partial \underset{\sim}{X}} = -\frac{\Omega\, \partial \delta\phi_g}{\partial \alpha} \quad . \qquad \text{(III.2.4)}$$

The R.H.S. of Eq. (III.2.1) can then be written as

$$(\frac{q}{m})[v_\parallel(\frac{\partial\delta\phi_g}{\partial \underset{\sim}{X}})(\frac{\partial}{\partial\varepsilon}) - \Omega\,\frac{\partial\delta\phi_g}{\partial\alpha}\,(\frac{\partial}{\partial\varepsilon} + \frac{\partial}{B_0\partial\mu})]f_{og} \quad . \tag{III.2.5}$$

In order to make contact with the low-frequency limit to be discussed later, we shall remove the $\partial/\partial\alpha$ terms in Eq. (III.2.5). Thus, we let

$$\delta f_g = (\frac{q}{m})\delta\phi_g(\frac{\partial}{\partial\varepsilon} + \frac{\partial}{B_0\partial\mu})f_{og} + \delta G_g \quad . \tag{III.2.6}$$

Equation (III.2.1) then reduces to

$$\pounds_g\delta G_g = -(\frac{q}{m})\{(\frac{\partial\delta\phi_g}{\partial t})(\frac{\partial}{\partial\varepsilon}) + [(\frac{\partial}{\partial t} + v_\parallel\,\frac{\partial}{\partial X_\parallel})\delta\phi_g](\frac{\partial}{B_0\partial\mu})\}f_{og} \quad . \tag{III.2.7}$$

Noting that

$$\delta G_g(\underset{\sim}{X},\mu,\varepsilon,\sigma,\alpha + 2\pi) = \delta G_g(\underset{\sim}{X},\mu,\varepsilon,\sigma,\alpha) \quad ; \tag{III.2.8}$$

i.e. δG_g is a periodic function in the gyrophase angle α, we let

$$\delta G_g = \sum_{n=-\infty}^{\infty} \langle\delta G_g\rangle_n\,\exp(-in\,\alpha) \quad , \tag{III.2.9}$$

where $\langle\delta G_g\rangle_n = \langle\delta G_g\rangle_n\,(\underset{\sim}{X},\,\mu,\,\varepsilon,\,\sigma)$. Equation (III.2.9) into Eq. (III.2.7), we find the following equation for $\langle\delta G_g\rangle_n$

$$\pounds_{gn}\langle\delta G_g\rangle_n \equiv (\frac{\partial}{\partial t} + v_\parallel\,\frac{\partial}{\partial x_\parallel} + in\Omega)\langle\delta G_g\rangle_n = -(\frac{q}{m})[(\frac{\partial f_{og}}{\partial\varepsilon})(\frac{\partial}{\partial t})$$

$$+ (\frac{\partial f_{og}}{B_0\partial\mu})(\frac{\partial}{\partial t} + v_\parallel\,\frac{\partial}{\partial X_\parallel})]\langle\delta\phi_g\rangle_n \quad . \tag{III.2.10}$$

Equation (III.2.10) is similar to the unmagnetized case and, hence, can be readily solved by Laplace-in-t and Fourier-in-$\underset{\sim}{X}$ transforms. Denoting

$$\delta\hat{A}_g = L_p(t)\, F_r\,(\underset{\sim}{X})\,[\delta A_g\,(\underset{\sim}{X},\, t)] \quad , \tag{III.2.11}$$

Eq. (III.2.10) yields

$$\langle\delta\hat{G}_g\rangle_n = \frac{-(q/m)[\omega(\partial f_{og}/\partial\varepsilon)+(\omega-k_\parallel v_\parallel)(\partial f_{og}/B_o\partial\mu)]}{\omega-k_\parallel v_\parallel-n\Omega}\,\langle\delta\hat{\phi}_g\rangle_n \quad . \tag{III.2.12}$$

Now, again, in order to make contact with later discussions on nonuniform plasmas where $v_\parallel$ is not a constant of motion due to varying $\underset{\sim}{B}_o$, we would like to remove the parallel (to $\underset{\sim}{B}_o$) propagator $\partial/\partial X_\parallel$ term from the R.H.S. of Eq. (III.2.7). Thus, we further let

$$\delta G_g = (-\,\frac{q}{m})\delta\phi_g\,\frac{\partial f_{og}}{B_o\partial\mu} + \delta h_g \quad . \tag{III.2.13}$$

Thus, $\langle\delta h_g\rangle_n$ satisfies

$$\pounds_{gn}\langle\delta h_g\rangle_n = -(\frac{q}{m})[(\frac{\partial f_{og}}{\partial\varepsilon})(\frac{\partial}{\partial t}) - in\,\Omega(\frac{\partial f_{og}}{B_o\partial\mu})]\langle\delta\phi_g\rangle_n \quad ; \tag{III.2.14}$$

or

$$\langle\delta\hat{h}_g\rangle_n = -\,\frac{q}{m}\cdot\frac{\omega\,\partial f_{og}/\partial\varepsilon + n\Omega\,\partial f_{og}/B_o\partial\mu}{\omega-k_\parallel v_\parallel-n\Omega}\,\langle\delta\hat{\phi}_g\rangle_n. \tag{III.2.15}$$

Summarizing the results obtained so far, we have

$$\delta\hat{f}_g = (\frac{q}{m})\delta\hat{\phi}_g\,\frac{\partial f_{og}}{\partial\varepsilon} + \sum_{n=-\infty}^{\infty}\langle\delta\hat{h}_g\rangle_n\,\exp\,(-in\alpha) \quad , \tag{III.2.16}$$

with $\langle\delta\hat{h}_g\rangle_n$ given by Eq. (III.2.15).

Since the field equations are in the $(\underset{\sim}{x}, t)$ space, we need to establish relations between $\delta\hat{A}_g$ and $\delta\hat{A} \equiv L_p(t)\, F_r(\underset{\sim}{x})[\delta A(\underset{\sim}{x},t)]$. Noting that $\delta A_g = \delta A$ we have

$$\delta A = L_p^{-1}(\omega)\, F_r^{-1}(\underset{\sim}{k})\, \delta\hat{A} = \int \frac{d\omega}{2\pi} \int \frac{d^3\underset{\sim}{k}}{(2\pi)^3}\, \delta\hat{A}(\omega,\underset{\sim}{k})\, \exp\,(-i\omega t + i\,\underset{\sim}{k}\cdot\underset{\sim}{x})$$

$$= L_p^{-1}(\omega) F_r^{-1}(\underset{\sim}{k})\delta\hat{A}_g = \int\frac{d\omega}{2\pi} \int \frac{d^3\underset{\sim}{k}}{(2\pi)^3}\, \delta\hat{A}_g(\omega,\underset{\sim}{k})\exp(-i\omega t + i\underset{\sim}{k}\cdot\underset{\sim}{X}) \quad . \tag{III.2.17}$$

Thus, noting $\underset{\sim}{X} = \underset{\sim}{X} + \underset{\sim}{v} \times \underset{\sim}{e}_\parallel/\Omega$, Eq. (III.1.6), we find

$$\delta\hat{A} = \delta\hat{A}_g\, \exp\,(iL_{\underset{\sim}{k}}) \quad , \tag{III.2.18}$$

where

$$L_{\underset{\sim}{k}} = \underset{\sim}{k}\cdot\frac{\underset{\sim}{v}\times\underset{\sim}{e}_\parallel}{\Omega} \quad . \tag{III.2.19}$$

Without any loss of generality, we may take $\underset{\sim}{k} = k_\perp\, \underset{\sim}{e}_1 + k_\parallel\, \underset{\sim}{e}_\parallel$. Thus, $L_{\underset{\sim}{k}} = \lambda \sin\alpha$ with $\lambda \equiv k_\perp\, v_\perp/\Omega$. Furthermore, we note the following identity

$$\exp(\pm i\lambda\, \sin\alpha) = \sum_{n=-\infty}^{\infty} J_n(\lambda)\exp(\pm\, in\alpha) \quad . \tag{III.2.20}$$

Hence,

$$\delta\hat{\phi}_g = \delta\hat{\phi}\, \exp(-i\lambda\sin\alpha) = \sum_n J_n(\lambda)\delta\hat{\phi}\, \exp(-in\alpha); \tag{III.2.21}$$

i.e.,

$$\langle \delta\hat{\phi}_g \rangle_n = \delta\hat{\phi}\, J_n(\lambda) \quad . \tag{III.2.22}$$

From Eqs. (III.2.18), (III.2.20), and (III.2.16), we obtain

$$\langle \delta\hat{f} \rangle_o \equiv \frac{1}{2\pi} \int_0^{2\pi} d\alpha\delta\hat{f} = \frac{q}{m}\, \delta\hat{\phi}\, [\frac{\partial f_o}{\partial \varepsilon} - \sum_n J_n^2 (\frac{\omega\partial/\partial\varepsilon + n\Omega\, \partial/B_o\partial\mu}{\omega - k_\parallel v_\parallel - n\Omega}) f_o] \quad . \tag{III.2.23}$$

Substituting Eq. (III.2.23) into the Poisson's equation then yields the following electrostatic dispersion relation

$$D_{e.s.} = 1 + \sum_j \chi_j = 0 \quad , \tag{III.2.24}$$

where χ_j, the jth-species susceptibility, is given by

$$\chi_j = -2\pi \frac{\omega_{pj}^2}{k^2} \sum_\sigma \int \frac{B_o d\varepsilon\, d\mu}{|v_\parallel|} [\frac{\partial f_o}{\partial\varepsilon} - \sum_n J_n^2 (\frac{\omega\partial/\partial\varepsilon + n\Omega\, \partial/B_o\partial\mu}{\omega - k_\parallel v_\parallel - n\Omega}) f_o] \quad , \tag{III.2.25}$$

$$= 2\pi \frac{\omega_{pj}^2}{k^2} \sum_\sigma \int \frac{B_o d\varepsilon d\mu}{|v_\parallel|} \sum_n J_n^2 (\frac{(k_\parallel v_\parallel + n\Omega)\partial/\partial\varepsilon + n\Omega\, \partial/B_o\partial\mu}{\omega - k_\parallel v_\parallel - n\Omega}) f_o \quad . \tag{III.2.26}$$

χ_j can be written in the more conventional form by transforming from the (ε,μ,σ) variables to the $(v_\perp, v_\parallel)$ variables. Noting that

$$\frac{\partial}{\partial\varepsilon} \rightarrow \frac{1}{v_\parallel} \frac{\partial}{\partial v_\parallel} \quad , \tag{III.2.27}$$

and

$$\frac{\partial}{B_o\partial\mu} \rightarrow \frac{\partial}{v_\perp \partial v_\perp} - \frac{1}{v_\parallel} \frac{\partial}{\partial v_\parallel} \quad ; \tag{III.2.28}$$

we have

$$\chi_j = 2\pi \frac{\omega_{pj}^2}{k^2} \int_{u_L} dv_\parallel \int_0^\infty v_\perp dv_\perp \sum_n J_n^2 \left(\frac{k_\parallel \partial/\partial v_\parallel + (n\Omega/v_\perp)\partial/\partial v_\perp}{(\omega - k_\parallel V_\parallel - n\Omega)}\right) f_o \quad . \qquad \text{(III.2.29)}$$

We conclude the discussion of this section by noting some identities involving Bessel functions; which have been used in deriving some of the above results and will be used in later derivations.

$$\sum_{n=-\infty}^{\infty} J_n^2 = 1 \quad , \qquad \text{(III.2.30)}$$

$$\sum_{n=-\infty}^{\infty} J_n J_n' = 0 \quad , \qquad \text{(III.2.31)}$$

$$\sum_{n=-\infty}^{\infty} n^2 J_n^2 = \frac{\lambda^2}{2} \quad , \qquad \text{(III.2.32)}$$

$$\sum_{n=-\infty}^{\infty} (J_n')^2 = \frac{1}{2} \quad . \qquad \text{(III.2.33)}$$

In addition, we have the familiar recurrence relations

$$J_{n+1} + J_{n-1} = \frac{2nJ_n}{\lambda} \quad , \qquad \text{(III.2.34)}$$

and

$$J_{n-1} - J_{n+1} = 2J_n' \quad . \qquad \text{(III.2.35)}$$

§III.3 Quasi-electrostatic and quasi-transverse approximations

In magnetized plasmas, the electrostatic (longitudinal $\delta\underset{\sim}{E}//\underset{\sim}{k}$) and electromagnetic (transverse $\delta\underset{\sim}{E} \perp \underset{\sim}{k}$) modes are usually coupled except in rather special limits; such as the exactly parallel propagation. The analysis, therefore, can be significantly simplified if we can assume the mode of interest is either essentially electrostatic or essentially electromagnetic. These so called quasi-electrostatic and quasi-transverse approximations are often adopted in plasma physics literature. The validity conditions quoted for the approximation are, however, usually qualitative. In this section, we will try to establish a formal procedure whereby one can quantitatively evaluate the validity limits of the two approximations.

From Eq. (I.1.7), we note that the wave equation is given as

$$\underset{\approx}{\varepsilon} \cdot \delta\hat{\underset{\sim}{E}} = \left(\underset{\sim}{k}\,\underset{\sim}{k} - k^2 \underset{\approx}{I} + \frac{\omega^2}{c^2} \underset{\approx}{D}\right) \cdot \delta\hat{\underset{\sim}{E}} = 0 \quad . \tag{III.3.1}$$

Now $\delta\hat{\underset{\sim}{E}}$ can be separated into its longitudinal and transverse components;

$$\delta\hat{\underset{\sim}{E}} = \delta\hat{E}_\ell\, \underset{\sim}{e}_k + \delta\hat{E}_t\, \underset{\sim}{e}_t \tag{III.3.2}$$

such that $\underset{\sim}{e}_t \cdot \underset{\sim}{e}_k = 0$. Dotting $\underset{\sim}{e}_k$ into Eq. (III.3.1), we have

$$\delta\hat{E}_\ell\; {}_kD_k + \delta\hat{E}_t\; {}_kD_t = 0 \quad , \tag{III.3.3}$$

where ${}_pD_q \equiv \underset{\sim}{e}_p \cdot \underset{\approx}{D} \cdot \underset{\sim}{e}_q$ for p,q = k,t. Meanwhile, dotting $\underset{\sim}{e}_t$ into Eq. (III.3.1) gives

$$\delta\hat{E}_t\,(n^2 - {}_tD_t) - \delta\hat{E}_\ell\; {}_tD_k = 0, \tag{III.3.4}$$

and $n = ck/\omega$.

For quasi-electrostatic approximation, we let $\omega = \omega_\ell + \delta\omega_\ell$ such that

$$_kD_k\,(\omega_\ell) = 0 \quad . \tag{III.3.5}$$

Equations (III.3.3) and (III.3.4) then become

$$\delta\hat{E}_\ell\,\delta\omega_\ell\,\frac{\partial}{\partial\omega_\ell}\,{}_kD_k{}^\ell + \delta\hat{E}_t\,{}_kD_t{}^\ell = 0 \quad , \tag{III.3.6}$$

and

$$\delta\hat{E}_t{}^\ell = \frac{\delta\hat{E}_\ell\,{}_tD_k^\ell}{\varepsilon_t^\ell} \quad . \tag{III.3.7}$$

Here, $\varepsilon_t = n^2 - {}_tD_t$ and $\varepsilon_t{}^\ell = \varepsilon_t(\omega_\ell)$. Equation (III.3.7) into Eq. (III.3.6), we obtain

$$\frac{\delta\omega_\ell}{\omega_\ell} = -\left[\frac{{}_tD_k\,{}_kD_t}{\varepsilon_t\,\partial({}_kD_k\omega_\ell)/\partial\omega_\ell}\right]^\ell \quad . \tag{III.3.8}$$

The quasi-electrostatic approximation is, thus, valid if

$$|\delta\omega_\ell/\omega_\ell| \ll 1 \quad . \tag{III.3.9}$$

Similarly, for the quasi-transverse approximation to be valid, we require $\omega = \omega_t + \delta\omega_t$ and $|\delta\omega_t/\omega_t| \ll 1$. Here, ω_t is given by

$$\varepsilon_t(\omega_t) = 0 \quad . \tag{III.3.10}$$

Following a similar procedure, the validity condition then becomes

$$\left|\frac{\delta\omega_t}{\omega_t}\right| = \left|\frac{{}_tD_k\ {}_kD_t}{{}_kD_k\ \partial(\omega_t\ \varepsilon_t)/\partial\omega_t}\right|^t << 1 \quad . \tag{III.3.11}$$

Noting that $\underset{\approx}{D} = \underset{\approx}{I} + i4\pi\underset{\approx}{\sigma}/\omega$, it is clear that, for $p\neq q$,

$$\begin{aligned} {}_pD_q &= \underset{\sim}{e}_p \cdot \left(\frac{i4\pi\underset{\approx}{\sigma}}{\omega}\right) \cdot \underset{\sim}{e}_q \\ &= \underset{\sim}{e}_p \cdot \frac{i4\pi}{\omega}\ \underset{\approx}{\sigma} \cdot \frac{\delta\hat{\underset{\sim}{E}}_q}{\delta\hat{E}_q} \quad . \end{aligned} \tag{III.3.12}$$

Thus, a physical explanation for the coupling between electrostatic (longitudinal) and electromagnetic (transverse) modes is that the current perturbation associated with one mode has a nonvanishing component when projected onto the polarization of the other mode. For example, if ${}_kD_t \neq 0$, then the current produced by $\delta\hat{\underset{\sim}{E}}_t$ has a finite component along $\underset{\sim}{k}$. With $\underset{\sim}{k} \cdot \delta\hat{\underset{\sim}{J}} \neq 0$ there will be charge density perturbations and, hence, space-charge (electrostatic) oscillations. Also, we remark that, since $\underset{\approx}{D} = \underset{\approx}{D}_h + i\underset{\approx}{D}_a$, this finite coupling can, sometimes, modify the wave growth or damping mechanism.

Let us illustrate the above discussions using the lower-hybrid wave; which is an electrostatic wave with $k_\perp >> k_\parallel$ and $\omega_{pi} \sim \omega << \omega_{ce}, \omega_{pe}$. The electrostatic dispersion relation is then given by

$${}_kD_k \simeq 1 - \frac{\omega_{pi}^2}{\omega^2} + \frac{\omega_{pe}^2}{\omega_{ce}^2} - \frac{\omega_{pe}^2 k_\parallel^2}{k^2\omega^2} = 0 \quad . \tag{III.3.13}$$

From the argument of associated current perturbations, it is easy to see that couplings to electromagnetic modes with $\underset{\sim}{e}_t$ nearly parallel to $\underset{\sim}{B}_0$ dominate.

Since $\underset{\sim}{e}_k \cdot \underset{\sim}{e}_\parallel = k_\parallel/k$, we have

$$ {}_tD_k \simeq {}_kD_t \simeq - \frac{\omega_{pe}^2}{\omega^2} \frac{k_\parallel}{k} \quad . \tag{III.3.14} $$

Meanwhile, we have

$$ \varepsilon_t^\ell \simeq \frac{(c^2k^2-\omega_{pe}^2)}{\omega^2} \simeq \frac{c^2k^2}{\omega^2} \quad ; \tag{III.3.15} $$

where we have assumed

$$ |ck|^2 \gg \omega_{pe}^2 \quad . \tag{III.3.16} $$

The validity criterion for the electrostatic approximation is then

$$ \left|\frac{\delta\omega_\ell}{\omega_\ell}\right| \simeq \frac{(\omega_{pe}k_\parallel/\omega k)^4}{2(1+\omega_{pe}^2/\omega_{ce}^2)} \left(\frac{\omega}{ck_\parallel}\right)^2 \sim n_\parallel^{-2} \ll 1 \quad . \tag{III.3.17} $$

Thus, Eqs. (III.3.16) and (III.3.17), which are consistent, constitute the validity conditions.

§III.4 Electromagnetic dielectric tensor for magnetized warm plasmas

The discussions given in Sections §III.1 and §III.2 can be straightforwardly generalized to allow full, electromagnetic perturbations. In this case, δf_g satisfies

$$ \mathcal{L}_g \delta f_g = - \frac{q}{m} \delta \underset{\sim}{a}_g \cdot \frac{\partial f_{og}}{\partial \underset{\sim}{V}} \ , \tag{III.4.1} $$

where

$$\delta \underset{\sim}{a}_g = \delta \underset{\sim}{E}_g + \frac{\underset{\sim}{v} \times \delta \underset{\sim}{B}_g}{c} \quad , \tag{III.4.2}$$

with $\pounds_g$ and $\partial/\partial \underset{\sim}{V}$ given, respectively by Eqs. (III.1.17) and (III.1.12). Again, noting $\delta\hat{f}_g = L_p F_r \delta f_g$ and letting

$$\delta\hat{f}_g = \sum_{n=-\infty}^{\infty} \langle \delta\hat{f}_g \rangle_n \exp(-in\ \alpha) \quad , \tag{III.4.3}$$

Eq. (III.4.1) becomes

$$\hat{\pounds}_{gn} \langle \delta\hat{f}_g \rangle_n = -\frac{q}{m} \langle \delta\hat{\underset{\sim}{a}}_g \cdot \frac{\partial}{\partial \underset{\sim}{V}} f_{og} \rangle_n \quad , \tag{III.4.4}$$

where

$$\hat{\pounds}_{gn} = -i(\omega - k_\parallel v_\parallel - n\Omega) \quad , \tag{III.4.5}$$

$$\langle \delta\hat{\underset{\sim}{a}}_g \cdot \frac{\partial}{\partial \underset{\sim}{V}} f_{og} \rangle_n = \left(\frac{v_\perp n J_n \delta\hat{E}_1}{\lambda}\right) P_1(f_{og}) + i v_\perp J_n' \delta\hat{E}_2\ P_1(f_{og}) + v_\parallel J_n \delta\hat{E}_\parallel P_2(f_{og}) \ , \tag{III.4.6}$$

$$P_1(f_{og}) = \left[\frac{\partial}{\partial \varepsilon} + \left(\frac{\omega - k_\parallel v_\parallel}{\omega}\right) \frac{\partial}{B_o \partial \mu}\right] f_{og} \quad , \tag{III.4.7}$$

and

$$P_2(f_{og}) = \left[\frac{\partial}{\partial \varepsilon} + \left(\frac{n\Omega}{\omega}\right) \frac{\partial}{B_o \partial \mu}\right] f_{og} \quad . \tag{III.4.8}$$

In deriving Eq. (III.4.6), we have noted that $\underset{\sim}{k}_\perp = k_\perp \underset{\sim}{e}_1$, $\underset{\sim}{e}_2 = \underset{\sim}{e}_\parallel \times \underset{\sim}{e}_1$,

$$(\underset{\sim}{v}_\perp \cdot \underset{\sim}{e}_1)\delta\hat{E}_g = \left(\frac{v_\perp}{\lambda}\right) \frac{i\partial\delta\hat{E}_g}{\partial\alpha} \quad , \tag{III.4.9}$$

and

$$(\underset{\sim}{v}_\perp \cdot \underset{\sim}{e}_2)\delta\hat{E}_g = iv_\perp \frac{\partial\delta\hat{E}_g}{\partial\lambda} \quad . \tag{III.4.10}$$

With $\delta\hat{f}_g$ and hence $\delta\hat{f}$ determined, we can then calculate the perturbed current density $\delta\underset{\sim}{J}$ and obtain the conductivity and dielectric tensors. For this purpose, let us note the following relations.

$$\langle \underset{\sim}{v}_\perp \cdot \underset{\sim}{e}_1 \, \delta\hat{f}\rangle_0 \equiv \left(\frac{1}{2\pi}\right) \int_0^{2\pi} d\alpha \; v_\perp \cos\alpha \; \delta\hat{f} = \sum_n v_\perp \, nJ_n \frac{\langle\delta\hat{f}_g\rangle_n}{\lambda} \quad , \tag{III.4.11}$$

$$\langle \underset{\sim}{v}_\perp \cdot \underset{\sim}{e}_2 \, \delta\hat{f}\rangle_0 = -iv_\perp \sum_n J_n' \, \langle\delta\hat{f}_g\rangle_n \quad , \tag{III.4.12}$$

and

$$\langle v_\parallel \, \delta\hat{f}\rangle_0 = v_\parallel \sum_n J_n \langle\delta\hat{f}_g\rangle_n \quad . \tag{III.4.13}$$

Denoting $\underset{\sim}{e}_3 = \underset{\sim}{e}_\parallel$,

$$[\ldots]_{n,1} = \int_{u_L} dv_\parallel \int_0^\infty v_\perp dv_\perp (\ldots) \frac{P_1(f_{og})}{\omega - k_\parallel v_\parallel - n\Omega} \quad , \tag{III.4.14}$$

and

$$[\ldots]_{n,2} = \int_{u_L} dv_\parallel \int_0^\infty v_\perp dv_\perp (\ldots) \frac{P_2(f_{og})}{\omega - k_\parallel v_\parallel - n\Omega} \quad ; \tag{III.4.15}$$

the elements of the conductivity tensor are then given by

$$\frac{i4\pi\sigma_{11}}{\omega} = \left(\frac{2\pi\omega_p^2}{\omega}\right) \sum_n \left[\left(\frac{nJ_n v_\perp}{\lambda}\right)^2\right]_{n,1} \quad , \qquad \text{(III.4.16)}$$

$$\frac{i4\pi\sigma_{12}}{\omega} = i\left(\frac{2\pi\omega_p^2}{\omega}\right) \sum_n \left(\frac{v_\perp^2 nJ_n' J_n}{\lambda}\right)_{n,1} \quad , \qquad \text{(III.4.17)}$$

$$\frac{i4\pi\sigma_{13}}{\omega} = \left(\frac{2\pi\omega_p^2}{\omega}\right) \sum_n \left(\frac{v_\parallel v_\perp nJ_n^2}{\lambda}\right)_{n,2} \quad , \qquad \text{(III.4.18)}$$

$$\sigma_{21} = -\sigma_{12} \quad , \qquad \text{(III.4.19)}$$

$$\frac{i4\pi\sigma_{22}}{\omega} = \left(\frac{2\pi\omega_p^2}{\omega}\right) \sum_n \left[(v_\perp J_n')^2\right]_{n,1} \quad , \qquad \text{(III.4.20)}$$

$$\frac{i4\pi\sigma_{23}}{\omega} = -i\left(\frac{2\pi\omega_p^2}{\omega}\right) \sum_n \left[v_\perp v_\parallel J_n J_n'\right]_{n,2} \quad , \qquad \text{(III.4.21)}$$

$$\frac{i4\pi\sigma_{31}}{\omega} = \left(\frac{2\pi\omega_p^2}{\omega}\right) \sum_n \left(\frac{nv_\parallel v_\perp J_n^2}{\lambda}\right)_{n,1} \quad , \qquad \text{(III.4.22)}$$

$$\frac{i4\pi\sigma_{32}}{\omega} = i\left(\frac{2\pi\omega_p^2}{\omega}\right) \sum_n \left(v_\parallel v_\perp J_n J_n'\right)_{n,1} \quad , \qquad \text{(III.4.23)}$$

and

$$\frac{i4\pi\sigma_{33}}{\omega} = \left(\frac{2\pi\omega_p^2}{\omega}\right) \sum_n \left[(v_\parallel J_n)^2\right]_{n,2} \quad . \qquad \text{(III.4.24)}$$

Noting $\underset{\approx}{D} = \underset{\approx}{I} + i4\pi\underset{\approx}{\sigma}/\omega$, we then have the complete electromagnetic dielectric tensor for magnetized plasmas including thermal kinetic effects. Finally, we remark that, for $\int d^3\underset{\sim}{v}\; v_\parallel\, f_o = 0$, we have $\sigma_{13}=\sigma_{31}$ and $\sigma_{32} = -\sigma_{23}$.

§III.5 Normal modes

In magnetized plasmas, there exists a variety of normal modes and, therefore, our treatment here will and has to be selective. The emphasis will be on thermal kinetic effects which are outside the scope of the usual cold fluid description. We will, however, first briefly review the cold fluid results and the readers are strongly recommended to familiarize themselves with the fluid theory via standard textbooks such as Krall and Trivelpiece. In the cold fluid theory, the normal modes are characterized by examining the two limiting cases; i.e., parallel and perpendicular propagations. The reason is just simplicity. For a general oblique propagation, the various modes are strongly coupled and discussions can easily become tedious and complicated. This, of course, becomes even more so in a Vlasov kinetic theory.

The cold plasma normal modes with $\underset{\sim}{k}$ parallel to $\underset{\sim}{B}_0$ is summarized in Fig. (III.5.1) where we have assumed the density is high ($\omega_{pe}>\omega_{ce}$).

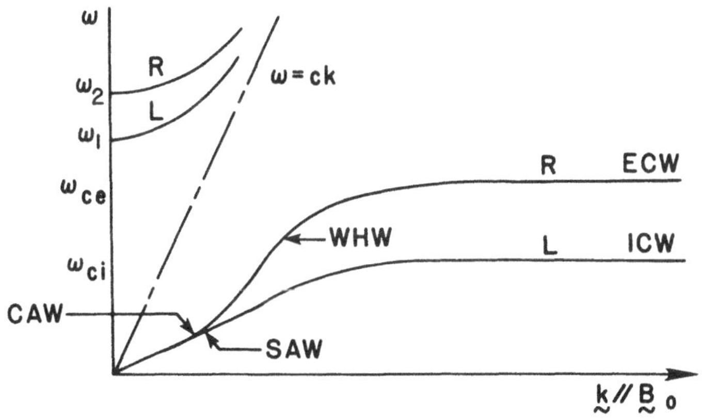

Fig. (III.5.1) Sketch of ω vs. k dispersion curves for waves with $\underset{\sim}{k}$ parallel to $\underset{\sim}{B}_0$ in a cold, magnetized, high-density ($\omega_{pe} > \omega_{ce}$) plasma.

In Fig. (III.5.1), R and L denote, respectively, right and left circular polarizations. ECW and ICW correspond to electron and ion cyclotron waves. CAW and SAW denote compressional and shear Alfvén waves, respectively. WHW denotes whistler wave and

$$\omega_{1,2} = \frac{\omega_{ce}}{2} \left[(1 + 4\, \omega_{pe}{}^2/\omega_{ce}{}^2)^{1/2} \mp 1 \right] \quad . \tag{III.5.1}$$

Figure (III.5.2) summarizes the cold plasma normal modes with $\underset{\sim}{k}$ perpendicular to $\underset{\sim}{B}_0$. Here, X and O denote, respectively, extraordinary and ordinary modes.

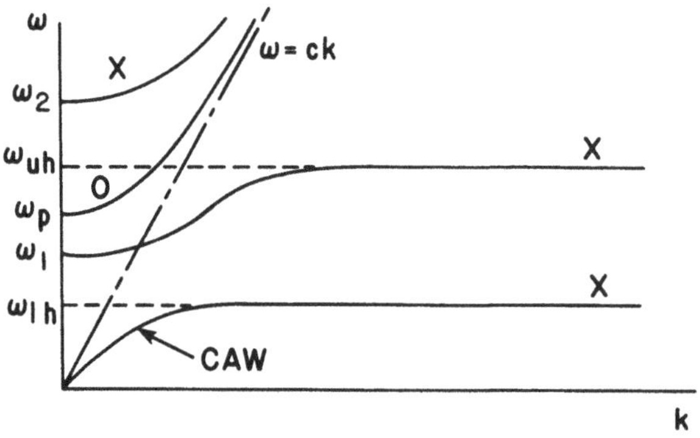

Fig. (III.5.2) Sketch of ω vs. k dispersion curves for waves with $\underset{\sim}{k}$ perpendicular to $\underset{\sim}{B}_0$ in a cold, magnetized high-density ($\omega_{pe} > \omega_{ce}$) plasma.

In Fig (III.5.2) ω_{Lh} and ω_{uh} corresponds to lower- and upper-hybrid frequencies given by

$$\omega_{Lh}{}^2 = \frac{\omega_{pi}^2}{1+\omega_{pe}^2/\omega_{ce}^2} = \frac{\omega_{ce}\omega_{ci}}{1+\omega_{ce}^2/\omega_{pe}^2} \quad , \tag{III.5.2}$$

and

$$\omega_{uh}^2 = \omega_{pe}^2 + \omega_{ce}^2 \quad . \tag{III.5.3}$$

Now, as we add the kinetic effects into the normal-mode description, two dominant features appear. One is the finite-Lamor-radius (FLR) effect; which, in addition to modifying the cold plasma modes, can give rise to new modes near the harmonics of the cyclotron frequencies; e.g. the Bernstein waves. The other feature is the wave-particle resonance; i.e., the cyclotron Landau resonance and the resonance condition is $\omega - k_\parallel v_\parallel - n\Omega = 0$. Thus, similar to the unmagnetized case, waves can be damped or excited by tapping the velocity-space free energy. It is worthy noting that finite plasma pressure can also give rise to new modes (e.g. the slow ion sound wave) as well as the so-called finite-β [$\beta = 8\pi N_o(T_e + T_i)/B_o^2$] effects which are responsible for certain types of instabilities. Since, in most cases, thermal effects can be included via the warm fluid description, we shall be concentrating in these lectures on the kinetic effects.

First, we illustrate the FLR physics by examining the ion Bernstein waves which are electrostatic waves with $k_\parallel = 0$. The corresponding dispersion relation is then given by

$$1 + \sum_j \chi_j = 0 \quad , \tag{III.5.4}$$

where

$$\chi_j = 2\pi \frac{\omega_{pj}^2}{k^2} \int_o^\infty v_\perp dv_\perp \sum_{n=1}^\infty J_n^2 \frac{2n^2\Omega_j^2}{\omega^2 - n^2\Omega_j^2} \frac{\partial f_{oj}}{\partial(v_\perp^2/2)} \quad . \tag{III.5.5}$$

If we further assume f_{oj} to be Maxwellian in $v_\perp$; i.e.,

$$f_{oj}(v_\perp) = (\pi v_{tj}^2)^{-1} \exp(-\frac{v^2}{v_{tj}}) \equiv f_{Mj} \quad , \tag{III.5.6}$$

and noting the following identity

$$2\pi \int_o^\infty v_\perp dv_\perp \, f_{Mj} \, J_n^2 = I_n(b_j)\exp(-b_j) \equiv \Gamma_{nj} \; ; \tag{III.5.7}$$

with $b_j = k_\perp^2 \rho_j^2/2$; χ_j reduces to

$$\chi_j = -\frac{1}{k^2\lambda_{Dj}^2} \sum_{n=1}^{\infty} \frac{2n^2\Omega_j^2}{\omega^2-n^2\Omega_j^2} \Gamma_{nj} \quad . \tag{III.5.8}$$

For ion Bernstein waves, we have $|\omega/\Omega_e| \sim O(\Omega_i/\Omega_e) << 1$ and $|k_\perp\rho_e \sim O(\rho_e/\rho_i)$ $<< 1$ and, hence, χ_e is dominated by the n=1 term or

$$\chi_e \simeq \frac{2\Gamma_{1e}}{k^2\lambda_{De}} \simeq \frac{\omega_{pe}^2}{\Omega_e^2} \quad . \tag{III.5.9}$$

The dispersion relation, Eq. (III.5.4), can then be written as

$$1 + \frac{\omega_{pe}^2}{\Omega_e^2} = \frac{1}{k^2\lambda_{Di}^2} \sum_{n=1}^{\infty} \frac{2n^2\Omega_i^2}{\omega^2-n^2\Omega_i^2} \Gamma_{ni} \equiv F_i(\omega) \quad . \tag{III.5.10}$$

Noting that $1 + \omega_{pe}^2/\Omega_e^2 \sim O(1)$, we can estimate ω by examining the limiting forms of $F_i(\omega)$. First, in the $|k| \to \infty$ limit, we have $|\Gamma_{ni}/k^2| \to 0^+$ and, thus,

$$\lim_{|k|\to\infty} \omega^2 - n^2\,\Omega_i^2 \to 0^+ \text{ for } n = 1, \ldots \quad . \tag{III.5.11}$$

In the long-wave length $k \to 0^+$ limit, noting that $\Gamma_{ni} \to (k^2)^n$, we find

$$\lim_{k\to 0^+} F_i(\omega) \to \frac{\omega_{pi}^2}{\omega^2-\Omega_i^2} + \sum_{n=2}^{\infty} \frac{2n^2\Omega_i^2}{\omega^2-n^2\Omega_i^2}\, O(|k^2|^{n-1}) \quad . \tag{III.5.12}$$

A sketch of Eq. (III.5.12) vs. ω/Ω_i is shown in Fig. (III.5.3).

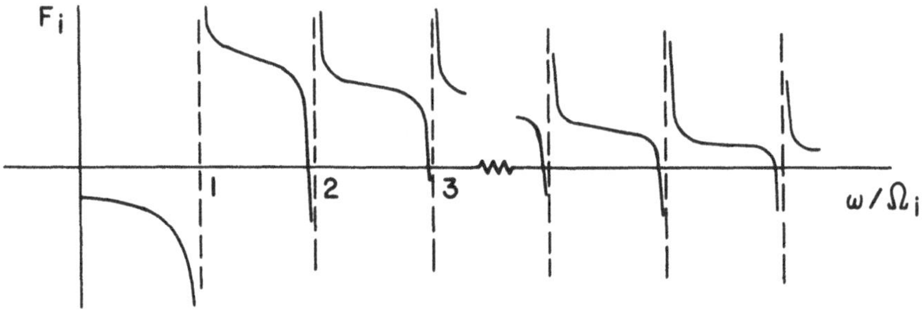

Fig. (III.5.3) Sketch of F_i given by Eq. (III.5.12) vs. ω/Ω_i.

From Eq. (III.5.12), it is clear that the "knee" in Fig. (III.5.3) is given by $\omega_{pi}^2/(\omega^2-\Omega_i^2)$. Equation (III.5.10) will thus produce roots at $n\Omega_i$ for $n \geq 2$ as well as at

$$1 + \frac{\omega_{pe}^2}{\Omega_e^2} \simeq \frac{\omega_{pe}^2}{\omega^2-\Omega_i^2} \qquad ; \qquad \text{(III.5.13)}$$

i.e., at $\omega^2 \simeq \omega_{lh}^2 = \omega_{pi}^2/[1 + (\omega_{pe}^2/\Omega_e^2)]$. From the above discussions on the two limiting cases, we can then sketch the ω - k dispersion curves as shown in Fig. (III.5.4).

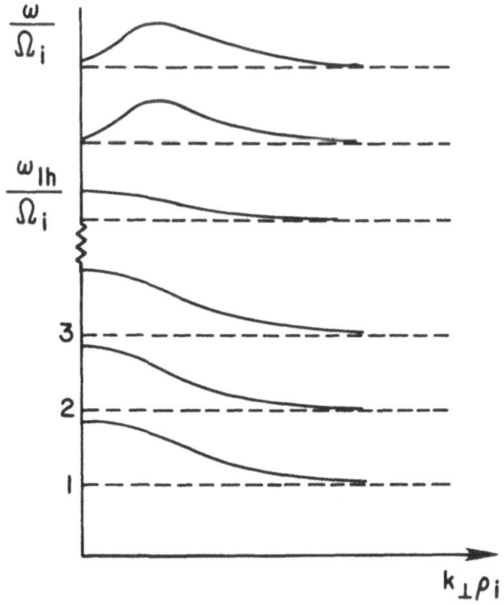

Fig. (III.5.4) Sketch of ω vs. $k_\perp \rho_i$ dispersion curves for ion Bernstein waves

In the next example, we shall consider the parallel-propagating electron and ion cyclotron waves including wave-particle resonances. From the cold fluid theory, we know that both waves are electromegnetic waves with $\underset{\sim}{k} \cdot \delta\hat{\underset{\sim}{E}} = 0$. since their phase velocities are, however, much less than the speed of light, we expect effects of wave-particle resonances to be important. To recover these waves from the Vlasov theory, we note that, as $k_\perp \to 0$, $J_o \to 1$, $J_{\pm 1}^2 \to \lambda^2/4$ and $J_{|n|\geq 2} \to 0$. Thus,

$$\frac{i4\pi\sigma_{11}}{\omega} \to \sum_j \left[\left(\frac{\pi\omega_{pj}^2}{2\omega}\right) \sum_{n=\pm 1} (v_\perp^2)_{n,1}\right] \equiv \sigma_+ + \sigma_- \equiv \sigma \; , \qquad \text{(III.5.14)}$$

$$\frac{i4\pi\sigma_{12}}{\omega} \to i\,(\sigma_+ - \sigma_-) \equiv \tilde{\sigma} \quad , \qquad \text{(III.5.15)}$$

and

$$\frac{i4\pi\sigma_{22}}{\omega} \to \sigma \quad . \qquad \text{(III.5.16)}$$

The normal-mode wave equation then becomes

$$(n^2 - 1 - \sigma)\, \delta\hat{E}_1 - \tilde{\sigma}\, \delta\hat{E}_2 = 0 \quad , \qquad \text{(III.5.17)}$$

and

$$\tilde{\sigma}\, \delta\hat{E}_1 + (n^2 - 1 - \sigma)\, \delta\hat{E}_2 = 0 \quad . \qquad \text{(III.5.18)}$$

We choose new basis for the eigenvectors $\delta\hat{E}_1$ and $\delta\hat{E}_2$ by letting

$$\delta\hat{E}_L = \delta\hat{E}_1 + i\, \delta\hat{E}_2 \quad , \qquad \text{(III.5.19)}$$

and

$$\delta\hat{E}_R = \delta\hat{E}_1 - i\,\delta\hat{E}_2 \quad . \tag{III.5.20}$$

Equations (III.5.17) and (III.5.18) become

$$\varepsilon_R\,\delta\hat{E}_R = 0 \quad , \tag{III.5.21}$$

and

$$\varepsilon_L\,\delta\hat{E}_L = 0 \quad ; \tag{III.5.22}$$

where

$$\varepsilon_R = n^2 - 1 - \sigma - i\,\tilde{\sigma}$$

$$= n^2 - 1 - 2\sigma_- \quad , \tag{III.5.23}$$

and

$$\varepsilon_L = n^2 - 1 - 2\sigma_+ \quad . \tag{III.5.24}$$

Noting that Ω carries charge sign, it is clear $\varepsilon_R=0$ corresponds to right hand circularly polarized waves and, hence, includes electron cyclotron Landau damping with the resonance condition given by $\omega - k_\parallel v_\parallel - |\Omega_e| = 0$. Similarly, for $\varepsilon_L = 0$, we have waves which resonant with ions and the resonance condition is $\omega - k_\parallel v_\parallel - \Omega_i = 0$. As a specific example, it is easy to show that, for the right hand circularly polarized wave and $|\omega| \gg |\Omega_i|$, the corresponding dispersion relation is given by

$$\frac{c^2k^2}{\omega^2} - 1 - \pi \frac{\omega_{pe}^2}{\omega} \int_{u_L} dv_\parallel \int_0^\infty v_\perp^3 \, dv_\perp \frac{[(\partial/v_\perp \partial v_\perp)-(k_\parallel v_\parallel/\omega)(\partial/v_\perp \partial v_\perp - \partial/v_\parallel \partial v_\parallel)]f_o}{\omega - k_\parallel v_\parallel - |\Omega_e|} = 0. \tag{III.5.25}$$

Equation (III.5.25) will be used in the next section to discuss electron cyclotron (whistler) instabilities driven by velocity-space anisotropy.

§III.6 Instabilities

In this section, we shall consider two instabilities due to anisotropic distributions. One is the whistler instability and the other is the ion loss-cone instability.

For the whistler instability, we let $\underset{\sim}{k}$ be parallel to $\underset{\sim}{B}_o = B_o \underset{\sim}{e}_z$; i.e., $\underset{\sim}{k} = k \underset{\sim}{e}_z$. We further simplify the analysis by taking

$$f_{oe}(\underset{\sim}{v}) = \delta(v_\parallel) f_M(v_\perp) \quad . \tag{III.6.1}$$

From Eq. (III.5.25), the whistler (ECW) dispersion relation, in this case, reduces to

$$\frac{c^2k^2}{\omega^2} - 1 + \frac{\omega_{pe}^2}{\omega} \frac{1}{(\omega-|\Omega_e|)} \left(1 + \frac{k^2 v_{t\perp}^2}{2\omega(\omega-|\Omega_e|)}\right) = 0 \ . \tag{III.6.2}$$

That Eq. (III.6.2) exhibits instabilities can be easily seen by taking the $k\to\infty$ limit and we find

$$\omega - |\Omega_e| = \frac{i\omega_{pe} v_{t\perp}}{\sqrt{2}c} \quad . \tag{III.6.3}$$

We note also that, from Eq. (III.6.2), it is clear that this whistler instability may be regarded as the extension of Weibel instability into a magnetized plasma.

Next, we consider the ion loss-cone instability. Assuming electrostatic pertubations and noting that $|k_\perp \rho_e|$, $|k_\parallel v_{et}/\omega| << 1$, we have

$$\chi_e \simeq - \omega_{pe}^2 k_\parallel^2/\omega^2 k^2 + \omega_{pe}^2/\Omega_e^2 \quad . \tag{III.6.4}$$

In the RHS of Eq. (III.6.4), we remark that the first term is due to electron parallel motion and the second term is due to electron polarization drift. Meanwhile, for ions, we have, from Eq. (III.2.29),

$$\chi_i = - \left(\frac{2\pi\omega_{pi}^2}{k^2}\right) \int dv_\parallel \int_0^\infty v_\perp dv_\perp \sum_n J_n^2 \left[P\left(\frac{1}{k_\parallel v_\parallel - \omega + n\Omega_i}\right) + i\pi\delta(k_\parallel v_\parallel - \omega + n\Omega_i)\right]$$

$$\times \left(k_\parallel \frac{\partial}{\partial v_\parallel} + \frac{n\Omega_i}{v_\perp} \frac{\partial}{\partial v_\perp}\right) f_{oi} \quad . \tag{III.6.5}$$

Now, since the instability is due to wave-particle (ion) resonance and noting that $|\Omega_i| >> |k_\parallel v_{i\parallel}|$, we, therefore, expect that the unstable modes occur very near the ion cyclotron harmonics. Let us assume $|\omega - m\Omega_i| \sim |k_\parallel v_{i\parallel}| << |\Omega_i|$. Thus,

$$\chi_i \simeq - \left(\frac{2\pi\omega_{pi}^2}{k^2}\right) \int dv_\parallel \int_0^\infty v_\perp dv_\perp J_m^2 \left[P\left(\frac{1}{k_\parallel v_\parallel - \omega + m\,\Omega_i}\right) + i\pi\delta(k_\parallel v_\parallel - \omega + m\Omega_i)\right]$$

$$\times \left(\frac{m\Omega_i}{v_\perp} \frac{\partial}{\partial v_\perp} f_{oi}\right) \quad . \tag{III.6.6}$$

To further simplify the analysis, we shall assume that

$$\omega = \omega_r + i\,\gamma = m\Omega_i + i\gamma \quad . \tag{III.6.7}$$

χ_i then reduces to

$$\chi_i \simeq -i\pi\left(\frac{2\pi\omega_{pi}^2}{k^2}\right)\left(\frac{m\Omega_i}{|k_\parallel|}\right) \int_0^\infty dv_\perp \; J_m^2 \left.\frac{\partial f_{oi}}{\partial v_\perp}\right|_{v_\parallel=0} \quad . \tag{III.6.8}$$

Now since $\partial(1 + \chi_e)/\partial\omega_r > 0$; i.e., the wave is of positive energy, we would have instabilities if

$$\int_0^\infty dv_\perp \; J_m^2 \left.\frac{\partial f_{oi}}{\partial v_\perp}\right|_{v_\parallel=0} > 0 \quad . \tag{III.6.9}$$

For ion loss-cone distributions and ignoring any effects due to the ambi-polar potential, it is reasonable to assume that

$$\left.\frac{\partial f_{oi}}{\partial v_\perp}\right|_{v_\parallel=0} > 0 \quad \text{for} \quad 0 < v_\perp < v_{\perp m} \quad . \tag{III.6.10}$$

Furthermore, for $|k_\perp v_\perp/\Omega_i| > m > 1$, we have

$$J_m(\lambda) \sim \left(\frac{2}{\pi\lambda}\right)^{1/2} \cos\left(\lambda - \frac{m\pi}{2} - \frac{\pi}{4}\right) \quad . \tag{III.6.11}$$

Therefore, if

$$\left|\frac{k_\perp v_{\perp m}}{\Omega_i}\right| \gg m > 1 \quad , \tag{III.6.12}$$

then the integral in Eq. (III.6.9) is dominated by the interval between 0 and $v_{\perp m}$ and we would have instabilities.

Mid-Term Take-Home Examination

1. Consider the electrostatic weak beam-plasma instability for $kv_o = \omega_{pe}$, i.e., the maximum unstable regime.

   (1.a) Estimate the amplitude of $\delta\hat{E}$ at which the linear theory breaks down.

   (1.b) Calculate the amplitude of $\delta\hat{E}$ at which the unstable (slow) wave begins to trap the cold beam.

   Are the two estimates consistent with each other?

2. Let us consider the right hand circularly polarized electron cyclotron wave ($\omega < |\Omega_e|$). Given $\delta\underset{\sim}{E} = \delta\underset{\sim}{\hat{E}}_R \exp(-i\omega t + ikz)$ where $\underset{\sim}{k} = k\underset{\sim}{e}_z$ and $\underset{\sim}{B}_0 = B_0\underset{\sim}{e}_z$,

   (2.a) show that, for the resonant particles $(\omega - k_\parallel v_\parallel - |\Omega_e| = 0)$, the linearization $(|\partial\delta f_1/\partial\underset{\sim}{v}| << |\partial f_0/\partial\underset{\sim}{v}|)$ breaks down for

   $$t \underset{\sim}{>} \left(\frac{q}{m}\,\delta\hat{E}_{eff}k\right)^{-1/2} \equiv t_{tr} \text{ where } \delta\hat{E}_{eff} = v_\perp\delta\hat{B}_R/c \quad ;$$

   (2.b) from the single particle picture, show that t_{tr} corresponds to the bounce time of a particle trapped by the $q\underset{\sim}{v}_\perp \times \delta\underset{\sim}{B}_R/c$ force in the parallel direction.

3. Electrostatic ion cyclotron waves are in the following parameter regime $|\omega| \sim |n\Omega_i| << |\Omega_e|$, $|k_\perp\rho_e| << 1$, $v_e >> |\omega/k_\parallel| >> v_i$, and $|k_\parallel/k_\perp| << 1$. Please plot the ω vs. $k_\perp$ dispersion curves.

4. The following wave equation models Bohm-Gross wave

$$\left(\delta^2 \frac{\partial^2}{\partial x^2} - \frac{\partial^2}{\partial t^2} - \omega_o^2\right)\delta\phi(x,t) = 0 \quad .$$

Here $\delta^2 << 1$. Solve $\delta\phi(x,t)$ subject to the initial perturbation that $\delta\phi(x,t = 0) = \exp(-x^2)$ and $\partial\delta\phi/\partial t = 0$ at $t = 0$. At least, try to establish the $t \to \infty$ behavior for finite fixed x and the $|x| \to \infty$ behavior for finite t. Please try to sketch your results.

5. Consider a one-dimensional unmagnetized plasma with $m_i \to \infty$. Now we impose an electrostatic potential given by

$$\phi_o = -\alpha x^2 \quad , \quad \alpha > 0$$

in order to electrostatically confine the plasma.

(5.a) Solve the unperturbed electron motion.

(5.b) Identify the constant of unperturbed motion.

(5.c) Use the constant of unperturbed motion to construct an equilibrium f_{oe} which is confined in space(i.e., $f_{oe} \to 0$ as $|x| \to \infty$).

(5.d) Solve, in as much detail as you can, the perturbed distribution, δf. In this case, it is suggested that you perform a transformation to the action-angle variables, where the action variable is the constant of motion derived in (5.b) and angle variable corresponds to the phase of oscillations in the ϕ_o potential well. Of course, you can do this problem in any way you like.

6. Give a physical explanation for the velocity-anisotropy driven R.H. polarized electron cyclotron wave. For this purpose, we shall assume $f_{oe}(\underset{\sim}{v}) = \delta(v_\parallel)\,\delta(v_\perp^2 - v_o^2)/\pi v_o^2$.

§III.7 Kinetic theory of low-frequency Magnetohydrodynamic waves

Magnetohydrodynamic (MHD) waves have frequencies, typically, much lower than the ion cyclotron frequency. Thus, they are capable of coupling to phenomena with time scales associated with plasma nonuniformities; such as diamagnetic drifts, magnetic trapping, resistive diffusion, and etc. We can then expect that MHD waves should play crucial roles in the physics of magnetically confined plasmas. Indeed, in the specific example of tokamaks, one finds ample such examples. For instances, we have the MHD stability analysis, anomalous transport due to drift waves, Alfvén wave heating and magnetic pumping. Since MHD waves are usually derived from the one-fluid description, it will be illuminating to consider these waves using the collisionless Vlasov description. This is one of the purposes of this section. The other purpose is to develop the low-frequency gyrokinetic formalism for uniform plasmas here; which can then be readily extended to nonuniform plasmas as we will do in the next chapter.

We first consider the electrostatic case in order to illustrate the theoretical approach. For low-frequency MHD waves, we adopt the following maximal orderings, with $\eta \ll 1$ being the smallness parameter,

$$\left|\frac{\omega}{\Omega}\right| \sim |k_\parallel \rho| \sim O(\eta) \quad \text{and} \quad |k_\perp \rho| \sim O(1) \quad . \tag{III.7.1}$$

From the results of Sec. §III.2; i.e., Eqs. (III.2.6) and (III.2.7), we have

$$\delta f_g = \frac{q}{m}\,\delta\phi_g\left(\frac{\partial}{\partial\varepsilon} + \frac{\partial}{B_o\partial\mu}\right)f_{og} + \delta G_g \quad , \tag{III.7.2}$$

and

$$\pounds_g\delta G_g = \left(\overset{\eta}{\frac{\partial}{\partial t}} + \overset{\eta}{v_\parallel\,\frac{\partial}{\partial X_\parallel}} - \overset{1}{\Omega\,\frac{\partial}{\partial\alpha}}\right)\delta G_g = -\frac{q}{m}\Big[\Big(\overset{\eta}{\frac{\partial}{\partial t}\,\delta\phi_g}\Big)\frac{\partial}{\partial\varepsilon}$$

$$+ \left(\frac{\partial}{\partial t} + \overset{\eta}{v_\parallel\,\frac{\partial}{\partial X_\parallel}}\right)\delta\phi_g\,\frac{1}{B_o}\,\frac{\partial}{\partial\mu}\Big]f_{og} \equiv \text{R.H.S.} \tag{III.7.3}$$

Here, in Eq. (III.7.3), we have also indicated the ordering of each term. We can now carry out a regular perturbative treatment. Let

$$\delta G_g = \delta G_{go} + \delta G_{g1} + \ldots \quad , \tag{III.7.4}$$

such that $|\delta G_{gn}/\delta G_{go}| \sim O(\eta^n)$. We have, for $O(1)$, $\delta G_{go}/\partial\alpha = 0$; i.e.,

$$\delta G_{go} = \delta G_{go}(\mu,\varepsilon,\underset{\sim}{X}) \tag{III.7.5}$$

To $O(\eta)$, we find

$$\left(\frac{\partial}{\partial t} + v_\parallel\,\frac{\partial}{\partial X_\parallel}\right)\delta G_{go} - \Omega\,\frac{\partial}{\partial\alpha}\,\delta G_{g1} = \text{R.H.S.} \tag{III.7.6}$$

Imposing the periodicity constraint to remove the secularity in α, we have

$$\left(\frac{\partial}{\partial t} + v_\parallel\,\frac{\partial}{\partial X_\parallel}\right)\delta G_{go} = \langle\text{R.H.S.}\rangle_0 \quad , \tag{III.7.7}$$

where

$$\langle \text{R.H.S.}\rangle_0 \equiv \frac{1}{2\pi}\int_0^{2\pi} d\alpha(\text{R.H.S.}) = -\frac{q}{m}\Big[\frac{\partial}{\partial t}\langle\delta\phi_g\rangle_0 \frac{\partial}{\partial t} + \Big(\frac{\partial}{\partial t} + v_\parallel \frac{\partial}{\partial X_\parallel}\Big)\langle\delta\phi_g\rangle_0$$

$$\times \frac{1}{B_0}\frac{\partial}{\partial\mu}\Big]f_{\underset{\sim}{v}og} \quad . \tag{III.7.8}$$

Equation (III.7.7) along with Eq. (III.7.8) is the low-frequency linear gyrokinetic equation for uniform magnetized plasmas.

To go further, we assume the perturbations have the plane-wave [i.e., $\exp(-i\omega t + i\underset{\sim}{k}\cdot\underset{\sim}{x})$] dependence. Noting discussions leading to Eq. (III.2.22), Eq. (III.7.7) then yields

$$\delta\hat{G}_{go} = -\frac{q}{m} J_0 \delta\hat{\phi}\Big(\frac{\omega}{\omega - k_\parallel v_\parallel}\frac{\partial}{\partial\varepsilon} + \frac{1}{B_0}\frac{\partial}{\partial\mu}\Big)f_0 \quad . \tag{III.7.9}$$

Meanwhile, Eq. (III.7.2) gives

$$\delta\hat{f} = \frac{q}{m}\delta\hat{\phi}\Big(\frac{\partial}{\partial\varepsilon} + \frac{\partial}{B_0\partial\mu}\Big)f_0 + \delta\hat{G}_{go}\exp(iL_k) + (\text{higher-order terms}) \quad . \tag{III.7.10}$$

Here, $L_k = \lambda \sin\alpha$. Substituting Eq. (III.7.10) into Poisson's equation yields the desired linear dispersion relation.

Let us consider the specific case with $f_0 = f_M$. It is then straightforward to derive, for j = species,

$$\delta\hat{n}_j = N_{oj}\oint d\alpha \int_0^\infty v_\perp dv_\perp \int_{u_L} dv_\parallel\, \delta\hat{f}_j = -N_{oj}\Big(\frac{q}{T}\Big)_j \delta\hat{\phi}(1 + \Gamma_{oj}\xi_j Z_j) \quad . \tag{III.7.11}$$

Here, $T_j = mv_{tj}^2/2$, $\Gamma_{oj} = I_0(b_j)\exp(-b_j)$, $b_j = k_\perp^2 v_{tj}^2/2\Omega_j^2$, $\xi_j = \omega/|k_\parallel v_{tj}|$, and $Z_j = Z(\xi_j)$. The corresponding dispersion relation is then given by

$$1 + \sum_j \chi_j = 0 \quad , \tag{III.7.12}$$

with

$$\chi_j = \frac{1}{k^2\lambda_{Dj}^2}\left(1 + \Gamma_{oj}\xi_j Z_j\right) \quad . \tag{III.7.13}$$

For the slow ion-sound wave, $\omega \sim k_\parallel c_s$, we assume $v_{ti} << \omega/k_\parallel << v_{te}$ and $k_\perp \rho_e << 1$. Thus, neglecting wave-particle resonance, we have

$$\chi_{er} \approx \frac{1}{k^2\lambda_{De}^2} \quad , \tag{III.7.14}$$

and

$$\chi_{ir} \simeq \left(\frac{1}{k^2\lambda_{Di}^2}\right)\left[1 - \Gamma_{oi}\left(1 + \frac{k_\parallel^2 v_{ti}^2}{2\omega^2}\right)\right] \quad . \tag{III.7.15}$$

Since $|k^2\lambda_{De}^2| << 1$, Eq. (III.7.12) becomes the quasi-neutrality condition and we obtain

$$\omega^2 = \frac{\Gamma_{oi} k_\parallel^2 c_s^2}{[1+(T_e/T_i)(1-\Gamma_{oi})]} \quad . \tag{III.7.16}$$

Equation (III.7.16), thus, corresponds to the slow ion-sound wave with FILR corrections. In the $k_\perp \to 0$ limit, one recovers the MHD result, i.e., $\omega^2 = k_\parallel^2 c_s^2$. It is interesting to note that, with FILR effects included, $\partial\omega/\partial k_\perp \neq 0$; i.e., the wave can now propagate across the magnetic field and this can lead to cross-field energy transport which is absent in the usual MHD description. Furthermore, noting that

$$\chi_{ei} \simeq \frac{1}{k^2\lambda_{De}^2}\sqrt{\pi}\,\xi_e \quad , \tag{III.7.17}$$

and

$$\chi_{ii} \simeq \frac{1}{k^2\lambda_{Di}^2} \Gamma_{oi}\sqrt{\pi}\, \xi_i \exp(-\xi_i^2) \qquad \text{(III.7.18)}$$

are both positive; the wave, thus, suffers both electron and ion Landau damping. Since $\omega \sim k_\parallel c_s$, negligible ion Landau damping then requires $|\xi_i| \gg 1$ or, equivalently, $T_e \gg T_i$. Thus, as an example, for normal tokamak plasmas with $T_e \sim T_i$, the slow ion-sound wave is heavily ion Landau damped and, therefore, is not a normal mode in its proper sense.

At this stage, it is perhaps instructive to recover the more familiar drift-kinetic equation by taking the $k_\perp\rho \to 0$ limit of the gyrokinetic equation. Now, in the $k_\perp\rho \to 0$ limit, we have $\delta G_g \to \delta G$; i.e., $\underset{\sim}{X} \to \underset{\sim}{x}$. Thus,

$$\delta f \to \frac{q}{m}\,\delta\phi\left(\frac{\partial}{\partial\varepsilon} + \frac{\partial}{B_o\partial\mu}\right)f_o + \delta G \quad , \qquad \text{(III.7.19)}$$

and

$$\left(\frac{\partial}{\partial t} + v_\parallel \frac{\partial}{\partial x_\parallel}\right)\delta G = -\frac{q}{m}\left[\frac{\partial\delta\phi}{\partial t}\frac{\partial}{\partial\varepsilon} + \left(\frac{\partial}{\partial t} + v_\parallel\frac{\partial}{\partial x_\parallel}\right)\delta\phi\,\frac{1}{B_o}\frac{\partial}{\partial\mu}\right]f_o \quad . \qquad \text{(III.7.20)}$$

Equation (III.7.19) into Eq. (III.7.20), we obtain

$$\left(\frac{\partial}{\partial t} + v_\parallel \frac{\partial}{\partial x_\parallel}\right)\delta f = \frac{q}{m}\, v_\parallel \frac{\partial\delta\phi}{\partial x_\parallel}\frac{\partial f_o}{\partial\varepsilon} = -\frac{q}{m}\,\delta E_\parallel \frac{\partial f_o}{\partial v_\parallel} \quad . \qquad \text{(III.7.21)}$$

Equation (III.7.21) is, of course, the desired linear drift-kinetic equation for uniform magnetized plasmas and, here, $\delta f = \delta f(\mu, v_\parallel, \underset{\sim}{x})$. As may be expected, we remark that the drift-kinetic equation can also be derived from the Vlasov equation via the $1/B_0$ expansion.

We now discuss the generalization of the above theoretical approach to include the full, electromagnetic perturbations. We are, thus, dealing with the collisionless description of Alfvén waves. From Eqs. (III.4.4) to (III.4.8), we have

$$\hat{\mathcal{L}}_{gn}\langle\delta\hat{f}_g\rangle_n = -\frac{q}{m}\langle\delta\hat{\underset{\sim}{a}}_g\cdot\frac{\partial}{\partial\underset{\sim}{V}}f_{og}\rangle_n \quad , \tag{III.4.4}$$

where

$$\hat{\mathcal{L}}_{gn} = \frac{\partial}{\partial t} + v_{\parallel}\frac{\partial}{\partial X_{\parallel}} + in\,\Omega \quad , \tag{III.7.22}$$

$$\langle\delta\hat{\underset{\sim}{a}}_g\cdot\frac{\partial}{\partial\underset{\sim}{V}}f_{og}\rangle_n = \left(\frac{nJ_n\Omega\delta\hat{E}_1}{k_\perp}\right)P_1(f_{og}) + iv_\perp J_n'\delta\hat{E}_2P_1(f_{og}) + v_{\parallel}J_n\delta\hat{E}_{\parallel}P_2(f_{og}) \quad , \tag{III.4.6}$$

$$P_1(f_{og}) = \left[\frac{\partial}{\partial t} + \left(1 - \frac{k_{\parallel}v_{\parallel}}{\omega}\right)\frac{\partial}{B_0\partial\mu}\right]f_{og} \quad , \tag{III.4.7}$$

and

$$P_2(f_{og}) = \left[\frac{\partial}{\partial t} + \left(\frac{n\Omega}{\omega}\right)\frac{\partial}{B_0\partial\mu}\right]f_{og} \quad . \tag{III.4.8}$$

Here, again, we have taken $\underset{\sim}{k} = k_\perp\underset{\sim}{e}_1 + k_{\parallel}\underset{\sim}{e}_{\parallel}$, $\underset{\sim}{e}_{\parallel} = \underset{\sim}{B}_0/B$ and $\underset{\sim}{e}_2 = \underset{\sim}{e}_{\parallel}\times\underset{\sim}{e}_1$. Noting Eqs. (III.4.3), (III.4.9) and (III.4.10), Eq. (III.4.4) can be, equivalently, rewritten as

$$\left(-i\omega + ik_{\parallel}v_{\parallel} - \Omega\frac{\partial}{\partial\alpha}\right)\delta\hat{f}_g = -\left(\frac{q}{m}\right)\left[\left(\frac{i\Omega}{k_\perp}\right)\left(\frac{\partial\delta\hat{E}_{g1}}{\partial\alpha}\right)P_1(f_{og})\right.$$

$$+ iv_\perp(\frac{\partial\delta\hat{E}_{g2}}{\partial\lambda})P_1(f_{og}) + v_\parallel\delta\hat{E}_{g\parallel}\frac{\partial f_{og}}{\partial\varepsilon} + i(\frac{v_\parallel\Omega}{\omega})(\frac{\partial\delta\hat{E}_{g\parallel}}{\partial\alpha})(\frac{\partial}{B_o\partial\mu})f_{og}] \quad . \qquad \text{(III.7.23)}$$

We now remove terms involving $\partial\delta\hat{E}_{\tilde{g}}/\partial\alpha$ in the R.H.S. of Eq. (III.7.23) by letting

$$\delta\hat{f}_g = \delta\hat{f}_{ga} + \delta\hat{G}_g \quad , \qquad \text{(III.7.24)}$$

where

$$\delta\hat{f}_{ga} = (\frac{q}{m})[i(\frac{\delta\hat{E}_{g1}}{k_\perp})P_1(f_{og}) + i(\frac{v_\parallel\delta\hat{E}_{g\parallel}}{\omega})(\frac{\partial}{B_o\partial\mu})f_{og}] \quad . \qquad \text{(III.7.25)}$$

$\delta\hat{G}_g$ then satisfies

$$(-i\omega + ik_\parallel v_\parallel - \Omega\frac{\partial}{\partial\alpha})\delta\hat{G}_g = -(\frac{q}{m})[(\omega - k_\parallel v_\parallel)(\frac{\delta\hat{E}_{g1}}{k_\perp})P_1(f_{og})$$

$$+ iv_\perp(\frac{\partial\delta\hat{E}_{g2}}{\partial\lambda})P_1(f_{og}) + (\omega - k_\parallel v_\parallel)(\frac{v_\parallel\delta\hat{E}_{g\parallel}}{\omega})(\frac{\partial f_{og}}{B_o\partial\mu}) + \frac{v_\parallel\delta\hat{E}_{g\parallel}\partial f_{og}}{\partial\varepsilon}] \equiv \text{R.H.S.}$$

(III.7.26)

Applying the orderings given by Eq. (III.7.1), we find, with $\delta\hat{G}_g = \delta\hat{G}_{go} + \delta\hat{G}_{g1} + \ldots$, that $\partial\delta\hat{G}_{go}/\partial\alpha = 0$ and

$$-i(\omega - k_\parallel v_\parallel)\delta\hat{G}_{go} = \langle\text{R.H.S.}\rangle_0 \quad . \qquad \text{(III.7.27)}$$

Equation (III.7.27), thus, represents the electromagnetic low-frequency linear gyrokinetic equation in uniform magnetized plasmas.

We now examine the specific example where $f_{og} = f_M$. Equation (III.7.27) then yields

$$\delta\hat{G}_{go} = i\,\frac{q}{T}\Big[J_o\,\frac{\delta\hat{E}_1}{k_\perp} + i\,\frac{v_\perp J_o'}{\omega-k_\parallel v_\parallel}\,\delta\hat{E}_2 + \frac{v_\parallel J_o}{\omega-k_\parallel v_\parallel}\,\delta\hat{E}_\parallel\Big]f_M \quad . \qquad \text{(III.7.28)}$$

That is,

$$\delta\hat{f} = -i\,\frac{q}{T}\Big[(1 - J_o e^{iL_k})\frac{\delta\hat{E}_1}{k_\perp} - i\,\frac{v_\perp J_o'}{\omega-k_\parallel v_\parallel}\,e^{iL_k}\delta\hat{E}_2 - \frac{v_\parallel J_o}{\omega-k_\parallel v_\parallel}\,e^{iL_k}\,\delta\hat{E}_\parallel\Big]f_M \quad . \qquad \text{(III.7.29)}$$

$\delta\hat{f}$ can then be substituted into the Maxwell's equations and yield the desired linear dispersion relation. For low-frequency waves, it is convenient to use the Poisson's equation; which, for $|\omega_{pi}/\Omega_i|^2$ or $|k\lambda_{De}|^{-2} >> 1$, is approximated by the quasi-neutrality condition and we have

$$\sum_j \Big(\frac{q^2}{T}\Big)_j\Big[(1 - \Gamma_o)\frac{\delta\hat{E}_1}{k_\perp} + \Big(\frac{i}{|k_\parallel|}\Big)\Big(\frac{k_\perp\rho}{2}\Big)\Big(\frac{\partial\Gamma_o}{\partial b}\Big)Z\delta\hat{E}_2 + \Gamma_o(1 + \xi Z)\frac{\delta\hat{E}_\parallel}{k_\parallel}\Big]_j = 0 \quad . \qquad \text{(III.7.30)}$$

Meanwhile, neglecting the space-charge effect in Ampere's law, we find, from the $\underset{\sim}{e}_2$ component,

$$-k^2\delta\hat{E}_2 + \frac{i4\pi\omega\delta\hat{J}_2}{c^2} = 0 \quad , \qquad \text{(III.7.31)}$$

and, from the $\underset{\sim}{e}_\parallel$ component,

$$k_\perp k_\parallel\delta\hat{E}_1 - k_\perp^2\delta\hat{E}_\parallel + \frac{i4\pi\omega\delta\hat{J}_\parallel}{c^2} = 0 \quad . \qquad \text{(III.7.32)}$$

Here, $\delta\hat{J}_2$ and $\delta\hat{J}_\parallel$ are given, respectively, by

$$\delta\hat{J}_2 \simeq \delta\hat{J}_{2o} = \sum_j \Big[N_o q \int 2\pi v_\perp dv_\perp \int dv_\parallel(-iv_\perp)J_o'\delta\hat{G}_{go}\Big]_j$$

$$= \sum_j \left(\frac{N_o q^2}{T}\right)_j 2\pi \int v_\perp dv_\perp v_\perp J'_{oj}\left[J_o\left(\frac{\delta\hat{E}_1}{k_\perp} - \frac{\delta\hat{E}_\parallel(1+\xi Z)}{k_\parallel}\right) - \frac{iv_\perp J'_o Z\delta\hat{E}_2}{k_\parallel v_t}\right]_j f_{Mj}(v_\perp) \quad , \tag{III.7.33}$$

and

$$\delta\hat{J}_\parallel \simeq \delta\hat{J}_{\parallel o} = -i \sum_j \left(\frac{N_o q^2}{T}\right)_j\left[\Gamma_o\omega(1 + \xi Z)\frac{\delta\hat{E}_\parallel}{k_\parallel^2} + \left(\frac{\delta\hat{E}_2}{k_\parallel}\right)\left(\frac{k_\perp \rho v_t}{2}\right)\left(\frac{\partial\Gamma_o}{\partial b}\right)(1 + \xi Z)\right]_j . \tag{III.7.34}$$

Equations (III.7.30) to (III.7.32) along with Eqs. (III.7.33) and (III.7.34), thus, completely determine the linear dynamics.

To illustrate the physics further, we assume $|k_\parallel/k_\perp| << 1$ and ignore the compressional Alfvén wave (i.e., $\delta\hat{E}_2 \approx \delta\hat{J}_2 \approx 0$). The quasi-neutrality condition, Eq. (III.7.30) then becomes

$$\left(\frac{\delta\hat{E}_1}{k_\perp}\right)k_\perp^2\rho_s^2 + \left(\frac{\delta\hat{E}_\parallel}{k_\parallel}\right)\left(1 - \frac{k_\parallel^2 c_s^2}{\omega^2}\right) = 0 \quad . \tag{III.7.35}$$

In deriving Eq. (III.7.35), we have taken $k_\perp^2\rho_e^2 << k_\perp^2\rho_i^2 < 1$, $v_{te} >> \omega/k_\parallel >> v_{ti}$ and neglected wave-particle resonances. Meanwhile,

$$i4\pi\delta\hat{J}_\parallel \simeq \left(\frac{1}{\lambda_{De}^2}\right)\left(\frac{\omega}{k_\parallel}\right)\left(\frac{\delta\hat{E}_\parallel}{k_\parallel}\right)\left(1 - \frac{k_\parallel^2 c_s^2}{\omega^2}\right) \quad . \tag{III.7.36}$$

The parallel Ampere's law, Eq. (III.7.32), then reduces to

$$\frac{\delta\hat{E}_1}{k_\perp} - \frac{\delta\hat{E}_\parallel}{k_\parallel} + \left(\frac{\delta\hat{E}_\parallel}{k_\parallel}\right)\left(\frac{\omega}{k_\parallel v_A}\right)^2(k_\perp\rho_s)^{-2}\left(1 - \frac{k_\parallel^2 c_s^2}{\omega^2}\right) = 0 \quad . \tag{III.7.37}$$

Combining Eqs. (III.7.35) and (III.7.37), we obtain the desired dispersion relation

$$(1 - \frac{\omega^2}{k_\parallel^2 v_A^2})(1 - \frac{k_\parallel^2 c_s^2}{\omega^2}) = -k_\perp^2 \rho_s^2 \quad . \tag{III.7.38}$$

Thus, due to the FLR effects, the shear Alfvén wave and slow ion-sound wave are coupled. For $\beta_e = c_s^2/v_A^2 << 1$, we find, for the slow ion-sound waves

$$\omega_s^2 \simeq \frac{k_\parallel^2 c_s^2}{(1+k_\perp^2 \rho_s^2)} \quad ; \tag{III.7.39}$$

which is the $k_\perp^2 \rho_i^2 < 1$ limit of Eq. (III.7.16). On the other hand, for the shear Alfvén waves, we find

$$\omega_A^2 \simeq k_\parallel^2 v_A^2 (1 + k_\perp^2 \rho_s^2) \quad . \tag{III.7.40}$$

Equation (III.7.40) is the dispersion relation for the, sometimes called, kinetic shear Alfvén wave (KSAW). We note that KSAW has finite perpendicular group velocity, $\partial\omega/\partial k_\perp \neq 0$, and plays important roles in shear Alfvén wave heating [Hasegawa and Chen, P.R.L. (1974) and Phys. Fluids (1976)] as well as Alpha-particle instabilities [Rutherford and Rosenbluth, P.R.L. (1974)]. Now, the Landau-damping effects can be included by retaining the imaginary part of Z function. For $1 >> \beta \gtrsim m_e/m_i$, we have $v_{te} \gtrsim \omega/k_\parallel \simeq v_A >> v_{ti}$ and, hence, KSAW suffers mainly electron Landau damping. Physically, this occurs because, for $k_\perp^2 \rho_s^2 \neq 0$, we have $\delta\hat{E}_\parallel \neq 0$ and particles can be accelerated or decelerated along $\underset{\sim}{B}_o$.

Finally, we remark, referring to Eq. (III.7.34), that finite $\delta\hat{E}_2$ can also lead to wave-particle resonance. Since $\delta\hat{E}_2 \propto \delta\hat{B}_\parallel$, this resonance, called transit-time damping, is due to acceleration (deceleration) of guiding centers <u>along</u> a varying magnetic field; just as the mirroring mechanism.

Homework #6

(1) Use the scalar and vector potentials, $\delta\phi$ and $\delta\underset{\sim}{A}$, such that

$$\delta\underset{\sim}{B} = \underset{\sim}{\nabla} \times \delta\underset{\sim}{A} \quad , \tag{H.6.1}$$

$$\delta\underset{\sim}{E} = -\left(\frac{\partial}{\partial\underset{\sim}{x}}\,\delta\phi + \frac{1}{c}\,\frac{\partial\delta\underset{\sim}{A}}{\partial t}\right) \quad , \tag{H.6.2}$$

and

$$\underset{\sim}{\nabla} \cdot \delta\underset{\sim}{A} = 0 \quad . \tag{H.6.3}$$

(1.a) show that the low-frequency response can be given as

$$\delta\hat{f}_g = \delta\hat{f}_{ga} + \delta\hat{G}_g \quad , \tag{H.6.4}$$

where

$$\delta\hat{f}_{ga} = \frac{q}{m}\left[\delta\hat{\phi}_g\,\frac{\partial}{\partial\varepsilon} + \left(\delta\hat{\phi}_g - \frac{v_\parallel \delta\hat{A}_{\parallel g}}{c}\right)\frac{1}{B_o}\,\frac{\partial}{\partial\mu}\right]f_{og} \quad , \tag{H.6.5}$$

and, with $\delta\hat{G}_g = \delta\hat{G}_{go} + \delta\hat{G}_{g1} \ldots$,

$$\langle \pounds_g\rangle_o\,\delta G_{go} = -\langle R_g\rangle_o \quad , \tag{H.6.6}$$

$$\langle \pounds_g\rangle_o = v_\parallel\,\frac{\partial}{\partial X_\parallel} + \frac{\partial}{\partial t} \quad , \tag{H.6.7}$$

$$\langle R_g\rangle_o = + \frac{q}{m}\left[\frac{\partial f_{og}}{\partial\varepsilon}\,\frac{\partial}{\partial t}\,\langle\delta\psi_g\rangle_o + \frac{1}{B_o}\,\frac{\partial f_{og}}{\partial\mu}\left(\frac{\partial}{\partial t} + v_\parallel\,\frac{\partial}{\partial X_\parallel}\right)\langle\delta\psi_g\rangle_o\right] \quad , \tag{H.6.8}$$

$$\langle \delta\psi_g \rangle_o = \langle \delta\phi_g - \underset{\sim}{v} \cdot \frac{\delta \underset{\sim}{A}_g}{c} \rangle_o = J_o\left(\delta\hat{\phi} - \frac{v_\parallel \delta\hat{A}_\parallel}{c}\right) - \frac{v_\perp J_o' \delta\hat{B}_\parallel}{k_\perp c} \quad . \qquad \text{(H.6.9)}$$

(1.b) Show that the Maxwell's equations are

$$k^2 \delta\hat{\phi} = 8\pi^2 \sum_j q \int \frac{Bd\mu d\varepsilon}{|v_\parallel|} \left\{ \frac{q}{m}\left[\delta\hat{\phi} \frac{\partial f_o}{\partial \varepsilon} + \left(\delta\hat{\phi} - \frac{v_\parallel \delta\hat{A}_\parallel}{c}\right)\frac{1}{B_o} \frac{\partial f_o}{\partial \mu}\right] + \delta\hat{G}_{go} J_o \right\} \quad , \qquad \text{(H.6.10)}$$

$$\left(k^2 - \frac{\omega^2}{c^2}\right)\delta\hat{A}_\parallel = \frac{8\pi^2}{c} \sum_j q \int \frac{Bd\mu d\varepsilon}{|v_\parallel|} v_\parallel \left[\frac{q}{m}\left(\delta\hat{\phi} - \frac{v_\parallel \delta\hat{A}_\parallel}{c}\right)\frac{\partial f_o}{B_o \partial \mu}\right] + \delta\hat{G}_{go} J_o \Big]$$

$$+ \frac{i\omega}{c} \frac{\partial}{\partial x_\parallel} \delta\hat{\phi} \quad , \qquad \text{(H.6.11)}$$

$$\left(k^2 - \frac{\omega^2}{c^2}\right)\delta\hat{B}_\parallel = \frac{8\pi^2}{c} k_\perp \sum_j q_j \int \frac{Bd\mu d\varepsilon}{|v_\parallel|} v_\perp \, \delta\hat{G}_{go} J_o' \quad . \qquad \text{(H.6.12)}$$

(2) Either from Problem (1) or your class notes, establish the respective small parameters for

(2.a) neglecting the compressional Alfvén wave.

(2.b) Weak coupling between the shear Alfvén wave and slow ion-sound wave.

CHAPTER IV

Linear Waves and Instabilities in Nonuniform Plasmas

§IV.1 Geometric optics and the ray equations

[Ref. S. Weinberg, Phys. Rev. 126, 1899 (1962); I. B. Bernstein, Phys. Fluids 18, 320 (1975)]

In Chapter I, we have analyzed, using two time and space scales, the propagation of a wave packet in a time-stationary and uniform dielectric medium. There, we have found that a wave packet (or quasi particle) propagates at the group velocity, $\underset{\sim}{v}_{gr}$, given by

$$\underset{\sim}{v}_{gr} = \frac{\partial\Omega(\underset{\sim}{k})}{\partial \underset{\sim}{k}} \quad . \tag{IV.1.1}$$

Here, $\Omega(\underset{\sim}{k})$ and $\underset{\sim}{k}$ corresponds, respectively, to the frequency and wave vector of the fast time and space variables. Now, if we relax the restriction and allow the medium to be time-varying and nonumiform, it can be expected that the two-time-and-space-scale approach is still applicable if the temporal and spatial variations of the dielectric are sufficiently slow when compared with Ω and $\underset{\sim}{k}$. This is the essence of geometric optics; that is, geometric optics deals with the propagation of wave packets in a dielectric medium which is slowly varying in both time and space. The corresponding equations for the wave packet propagation are called the ray equations. Before we get into the detailed analysis, let us consider this subject from the quasi-particle point of view. In this perspective, the wave packet is characterized by Ω (energy) and $\underset{\sim}{k}$ (momentum). Denoting $\underset{\sim}{x}_1$ and t_1 as the slow space and time variables, we then have

$$\Omega = \Omega(\underset{\sim}{k}, \underset{\sim}{x}_1, t_1) \quad , \tag{IV.1.2}$$

and

$$\underset{\sim}{k} = \underset{\sim}{k}(\underset{\sim}{x}_1, t_1) \quad . \tag{IV.1.3}$$

Thus, the quasi-particle dynamics is completely specified if $d\underset{\sim}{x}_1/dt_1$, $d\Omega/dt_1$, and $d\underset{\sim}{k}/dt_1$ are known and those consitute the ray equations. Now, if variations in the medium are sufficiently slow, we then know that, at least in the lowest order, Eq. (IV.1.1) must still hold; i.e.

$$\frac{d\underset{\sim}{x}_1}{dt_1} = \frac{\partial\Omega}{\partial\underset{\sim}{k}} \quad . \tag{IV.1.4}$$

With $d\underset{\sim}{x}_1/dt_1$ given, we may move forward and derive the rest of the ray equations. In order to illustrate the theoretical approach, we first consider the case with the medium being time-stationary and, hence, shall employ the two space scales, $\underset{\sim}{x}_0$ and $\underset{\sim}{x}_1$. We now assume the corresponding wave equation is given as follows

$$\underset{\approx}{\varepsilon}(\underset{\sim}{x}_1, -i\frac{\partial}{\partial\underset{\sim}{x}}, i\frac{\partial}{\partial t}) \cdot \delta\underset{\sim}{E}(\underset{\sim}{x}_0, \underset{\sim}{x}_1, t) = 0 \quad . \tag{IV.1.5}$$

Letting

$$\delta\underset{\sim}{E}(\underset{\sim}{x}_0, \underset{\sim}{x}_1, t) = \delta\hat{\underset{\sim}{E}} \exp[iS(\underset{\sim}{x}_0, \underset{\sim}{x}_1, t)] \quad ; \tag{IV.1.6}$$

where

$$S = \underset{\sim}{k}(\underset{\sim}{x}_1) \cdot \underset{\sim}{x}_0 - \omega t \qquad \text{(IV.1.7)}$$

is the eikonal for this case. Equation (IV.1.5) then yields, in the lowest order,

$$\underset{\approx}{\varepsilon}(\underset{\sim}{x}_1, \underset{\sim}{k}, \omega) \cdot \delta\hat{\underset{\sim}{E}} = 0 \quad . \qquad \text{(IV.1.8)}$$

Here, we have noted that $\partial/\partial\underset{\sim}{x} = \partial/\partial\underset{\sim}{x}_0 + \partial/\partial\underset{\sim}{x}_1 \simeq \partial/\partial x_0$. Nontrivial solutions of $\delta\hat{\underset{\sim}{E}}$ give the following local dispersion relation

$$\varepsilon(\underset{\sim}{x}_1, \underset{\sim}{k}, \omega) = \|\underset{\approx}{\varepsilon}\| = 0 \quad . \qquad \text{(IV.1.9)}$$

Equivalently, Eq. (IV.1.9) can be solved for ω; i.e.,

$$\omega = \Omega(\underset{\sim}{x}_1, \underset{\sim}{k}) \quad . \qquad \text{(IV.1.10)}$$

Now, since $\varepsilon = 0$ is always satisfied along the ray trajectory, it is a constant of the motion; i.e.,

$$\dot{\varepsilon} \equiv \frac{d\varepsilon}{dt_1} = \frac{\partial\varepsilon}{\partial t_1} + \frac{\partial\varepsilon}{\partial\underset{\sim}{x}_1} \cdot \dot{\underset{\sim}{x}}_1 + \frac{\partial\varepsilon}{\partial\underset{\sim}{k}} \cdot \dot{\underset{\sim}{k}} + \frac{\partial\varepsilon}{\partial\omega}\dot{\omega} = 0 \quad . \qquad \text{(IV.1.11)}$$

With the medium being time-stationary, we have $\partial\varepsilon/\partial t_1 = 0$ and ω is a constant; $\dot{\omega} = 0$. Equation (IV.1.11), thus, becomes

$$\frac{\partial\varepsilon}{\partial\underset{\sim}{x}_1} \cdot \dot{\underset{\sim}{x}}_1 + \frac{\partial\varepsilon}{\partial\underset{\sim}{k}} \cdot \dot{\underset{\sim}{k}} = 0 \quad . \qquad \text{(IV.1.12)}$$

Noting Eq. (IV.1.4), we then find

$$\dot{k}_i = - \frac{(\partial\varepsilon/\partial x_{1j})(\partial\Omega/\partial k_j)}{(\partial\varepsilon/\partial k_i)} = \frac{(\partial\varepsilon/\partial x_{1i})_{\underset{\sim}{k},\Omega}}{(\partial\varepsilon/\partial\Omega\ )_{\underset{\sim}{k},\underset{\sim}{x}_1}} = - \frac{\partial\Omega}{\partial x_{1i}}\bigg|_{\underset{\sim}{k}} \quad . \tag{IV.1.13}$$

We note that, in the present case, Eq. (IV.1.13) can, of course, be derived more readily by using the fact that $\dot{\omega} = \dot{\Omega} = 0$.

Let us now generalize the above eikonal analysis to a medium varying in both time and space. The approach is somewhat different in the sense that no specific two scales are assumed. Thus, let the wave equation be

$$\underset{\approx}{\varepsilon}\left(\underset{\sim}{x}, t; - i \frac{\partial}{\partial \underset{\sim}{x}}, i \frac{\partial}{\partial t}\right) \cdot \delta\underset{\sim}{E}(\underset{\sim}{x}, t) = 0 \tag{IV.1.14}$$

Again, adopting the eikonal representation such that

$$\delta\underset{\sim}{E}(\underset{\sim}{x}, t) = \delta\hat{\underset{\sim}{E}} \exp[i\phi(\underset{\sim}{x}, t)] \quad , \tag{IV.1.15}$$

the local dispersion relation is then given by

$$\varepsilon(\underset{\sim}{x}, t ; \underset{\sim}{k}, \omega) \equiv \| \underset{\approx}{\varepsilon} \| = 0 \quad . \tag{IV.1.16}$$

Here, however, $\underset{\sim}{k}$ and ω must be properly defined as

$$\underset{\sim}{k} = \frac{\partial\phi}{\partial \underset{\sim}{x}}\bigg|_t = \underset{\sim}{k}(\underset{\sim}{x}, t) \quad , \tag{IV.1.17}$$

and

$$\omega = -\left.\frac{\partial\phi}{\partial t}\right|_{\mathbf{x}} = \omega(\mathbf{x},\ t) \quad . \qquad \text{(IV.1.18)}$$

Equation (IV.1.16), of course, implies two scales, i.e.,

$$\left|\frac{\partial \mathbf{k}}{\partial \mathbf{x}}\right| << |\mathbf{k}|^2 \quad , \quad \left|\frac{\partial\omega}{\partial t}\right| << |\omega|^2 \quad ,$$

$$\left|\frac{\partial \mathbf{k}}{\partial t}\right| << |\omega\mathbf{k}| \quad , \quad \left|\frac{\partial\omega}{\partial \mathbf{x}}\right| << |\omega\mathbf{k}| \quad . \qquad \text{(IV.1.19)}$$

Again, from Eq. (IV.1.16), we can obtain

$$\omega = \Omega(\mathbf{k};\ \mathbf{x},\ t) \quad . \qquad \text{(IV.1.20)}$$

We now proceed to derive the ray equations. First, we have, from Eq. (IV.1.17),

$$\dot{\mathbf{k}} = \left.\frac{\partial \mathbf{k}}{\partial t}\right|_{\mathbf{x}} + \left.\frac{\partial \mathbf{k}}{\partial \mathbf{x}}\right|_t \cdot \dot{\mathbf{x}} = \left.\frac{\partial \mathbf{k}}{\partial t}\right|_{\mathbf{x}} + \left.\frac{\partial \mathbf{k}}{\partial \mathbf{x}}\right|_t \cdot \left.\frac{\partial\Omega}{\partial \mathbf{k}}\right|_{\mathbf{x},t} \quad . \qquad \text{(IV.1.21)}$$

Meanwhile, noting Eqs. (IV.1.17), (IV.1.18), and (IV.1.20), we find

$$\left.\frac{\partial \mathbf{k}}{\partial t}\right|_{\mathbf{x}} = \frac{\partial^2\phi}{\partial t\partial \mathbf{x}} = -\left.\frac{\partial\omega}{\partial \mathbf{x}}\right|_t = -\left.\frac{\partial\Omega}{\partial \mathbf{k}}\right|_{\mathbf{x},t} \cdot \left.\frac{\partial \mathbf{k}}{\partial \mathbf{x}}\right|_t - \left.\frac{\partial\Omega}{\partial \mathbf{x}}\right|_{\mathbf{k},t} \quad . \qquad \text{(IV.1.22)}$$

Substituting Eq. (IV.1.22) into Eq. (IV.1.21), we derive

$$\dot{\mathbf{k}} = -\left.\frac{\partial\Omega}{\partial \mathbf{x}}\right|_{\mathbf{k},t} \quad . \qquad \text{(IV.1.23)}$$

For $d\omega/dt$, we find

$$\dot{\omega} = \dot{\Omega} = \frac{\partial\Omega}{\partial t} + \frac{\partial\Omega}{\partial \underset{\sim}{k}} \cdot \dot{\underset{\sim}{k}} + \frac{\partial\Omega}{\partial \underset{\sim}{x}} \cdot \dot{\underset{\sim}{x}} \quad . \tag{IV.1.24}$$

Noting that $\dot{\underset{\sim}{x}} = \partial\Omega/\partial\underset{\sim}{k}$ and Eq. (IV.1.23), we readily derive

$$\dot{\omega} = \dot{\Omega} = \left.\frac{\partial\Omega}{\partial t}\right|_{\underset{\sim}{k},\underset{\sim}{x}} \quad . \tag{IV.1.25}$$

Summarizing the results, the ray equations are

$$\dot{\underset{\sim}{x}} = \frac{d\underset{\sim}{x}}{dt} = \frac{\partial\Omega}{\partial\underset{\sim}{k}} \quad , \tag{IV.1.24}$$

$$\dot{\underset{\sim}{k}} = -\frac{\partial\Omega}{\partial\underset{\sim}{x}} \quad , \tag{IV.1.23}$$

$$\dot{\Omega} = \frac{\partial\Omega}{\partial t} \quad , \tag{IV.1.25}$$

where

$$\Omega = \Omega\left(\underset{\sim}{k}\ ;\ \underset{\sim}{x},\ t\right) \tag{IV.1.20}$$

satisfies the local dispersion relation

$$\varepsilon(\underset{\sim}{x},\ t;\ \underset{\sim}{k},\ \Omega) = 0 \quad . \tag{IV.1.16}$$

Looking at these ray equations from the quasi-particle picture, we may regard them as Hamilton equations of motion for the quasi particles with Hamiltonian $H = \Omega$, momentum $\underset{\sim}{p} = \underset{\sim}{k}$ and $\underset{\sim}{q} = \underset{\sim}{x}$. Furthermore, since the above ray equations conserve the number of quasi particles, they are strictly valid for a Hermitian dielectric. Extensions to include a small anti-Hermitian part will be considered in the next section.

Finally, we remark that geometric optics plays an important role in plasma physics because it serves as the first step toward our understanding of wave dynamics in realistic plasmas. For example, in radio-frequency heating of plasmas, ray tracing is routinely calculated in order to understand how the wave penetrates and, thereby, deposits its energy inside the plasma.

§IV.2 Derivation of the wave (quasi-particle) kinetic equation.

In Sec. §IV.1, we have shown that wave packets obey Hamilton equations of motion as quasi particles with $H = \Omega$ and $\underset{\sim}{p} = \underset{\sim}{k}$. It is, therefore, expected that, if we have a collection of wave packets, the corresponding distribution function, $N = N(\underset{\sim}{k},\underset{\sim}{x},t)$, would obey the following Liouville's equation

$$\frac{dN}{dt} = \frac{\partial N}{\partial t} + \dot{\underset{\sim}{x}} \cdot \frac{\partial N}{\partial \underset{\sim}{x}} + \dot{\underset{\sim}{k}} \cdot \frac{\partial N}{\partial \underset{\sim}{k}} = \frac{\partial N}{\partial t} + \frac{\partial \Omega}{\partial \underset{\sim}{k}} \cdot \frac{\partial N}{\partial \underset{\sim}{x}} - \frac{\partial \Omega}{\partial \underset{\sim}{x}} \cdot \frac{\partial N}{\partial \underset{\sim}{k}} = 0 \quad ; \qquad \text{(IV.2.1)}$$

which is called the wave kinetic equation in plasma physics literatures. Now, as we also mentioned in Sec. §IV.1, Eq. (IV.2.1) is valid only for a Hermitian dielectric in which there is no wave growth or damping such that $\dot{N} = 0$. In this section, we shall generalize Eq. (IV.2.1) to include a weak wave dissipation or growth. In another perspective, this discussion can also be regarded as extensions of our analysis in Sec. §I.3 to nonuniform media.

To simplify the presentation, we shall consider the one-dimensional case and let the wave equation be

$$D\left(x,t \; ; \; - i \frac{\partial}{\partial x} \; , \; i \frac{\partial}{\partial t}\right)\delta E(x,t) = 0 \quad , \qquad \text{(IV.2.2)}$$

where

$$D = \sum_{n=0}^{\infty} \sum_{m=0}^{\infty} f_{n,m}(x,t)(- i \frac{\partial}{\partial x})^n (i \frac{\partial}{\partial t})^m \quad . \tag{IV.2.3}$$

Thus, we have assumed D is analytic in $(\partial/\partial x)$ and $(\partial/\partial t)$; which, of course, is physically reasonable. We, furthermore, assume the existence of two time and space scales; i.e., we let (x_o,t_o) be fast variables and (x_1,t_1) the slow variables. Now, since the dielectric medium is slowly varying in space and time, we let $x = x_1 + x_o$ and $t = t_1 + t_o$ in D. Meanwhile, we Fourier transform δE in the (x_o,t_o) variables by letting

$$\delta E(x,t) = \delta E(x_o,t_o \ ; \ x_1,t_1) = \int dk \int d\omega \ \delta\hat{E}(k,\omega \ ; \ x_1, \ t_1) \exp(iS_o) \quad , \tag{IV.2.4}$$

with $s_o = kx_o - \omega t_o$. Equation (IV.2.2) then becomes

$$D(x_1 + x_o, \ t_1 + t_o \ ; \ - i \frac{\partial}{\partial x_o} - i \frac{\partial}{\partial x_1} \ , \ i \frac{\partial}{\partial t_o} + i \frac{\partial}{\partial t_1})\delta E(x,t)$$

$$\simeq \int dk \int d\omega [D(x_1, \ t_1 \ ; \ k,\omega) + x_o \frac{\partial D}{\partial x_1} + t_o \frac{\partial D}{\partial t_1}$$

$$+ \frac{\partial D}{\partial k} (- i \frac{\partial}{\partial x_1}) + \frac{\partial D}{\partial \omega} (i \frac{\partial}{\partial t_1})] \delta\hat{E} \exp(iS_o)$$

$$\simeq \int dk \int d\omega \ e^{iS_o} [(D_r + iD_i)\delta\hat{E} + i \frac{\partial}{\partial k}(\frac{\partial D_r}{\partial x_1} \delta\hat{E})$$

$$- i \frac{\partial}{\partial \omega} (\frac{\partial D_r}{\partial t_1} \delta\hat{E}) - i \frac{\partial D_r}{\partial k} \frac{\partial \delta\hat{E}}{\partial x_1} + i \frac{\partial D_r}{\partial \omega} \frac{\partial \delta\hat{E}}{\partial t_1}] = 0 \quad . \tag{IV.2.5}$$

In the zeroth order, Eq. (IV.2.5) gives

$$D_r |\delta\hat{E}|^2 = 0 \quad , \tag{IV.2.6}$$

Thus

$$|\delta\hat{E}|^2 = P(k;\, x_1,\, t_1)\delta(\omega - \Omega) \qquad , \tag{IV.2.7}$$

with Ω satisfying the local Hermitian dispersion relation

$$D_r(x_1,\, t_1;\, k,\Omega) = 0 \qquad ; \tag{IV.2.8}$$

and $P = \int d\omega|\delta\hat{E}|^2$. In the next order, we obtain

$$\frac{\partial D_r}{\partial \omega}\frac{\partial \delta\hat{E}}{\partial t_1} - \frac{\partial D_r}{\partial k}\frac{\partial \delta\hat{E}}{\partial x_1} + \frac{\partial}{\partial k}\left(\frac{\partial D_r}{\partial x_1}\delta\hat{E}\right) - \frac{\partial}{\partial \omega}\left(\frac{\partial D_r}{\partial t_1}\delta\hat{E}\right) = -D_i\delta\hat{E} \qquad . \tag{IV.2.9}$$

Performing $\delta\hat{E}^* \times$ Eq. (IV.2.9) + $\delta\hat{E} \times$ Eq. (IV.2.9)* yields

$$\frac{\partial D_r}{\partial \omega}\frac{\partial}{\partial t_1}|\delta\hat{E}|^2 - \frac{\partial D_r}{\partial k}\frac{\partial}{\partial x_1}|\delta\hat{E}|^2 + \frac{\partial}{\partial k}\left(\frac{\partial D_r}{\partial x_1}|\delta\hat{E}|^2\right) - \frac{\partial}{\partial \omega}\left(\frac{\partial D_r}{\partial t_1}|\delta\hat{E}|^2\right) = -2D_i|\delta\hat{E}|^2. \tag{IV.2.10}$$

For the sake of simplicity, let us assume

$$\frac{\partial^2 D_r}{\partial \omega \partial t_1} = \frac{\partial^2 D_r}{\partial k \partial x_1} = 0 \qquad . \tag{IV.2.11}$$

Equation (IV.2.10), integrated over ω, then becomes, noting Eq. (IV.2.7),

$$\frac{dN}{dt} = \left(\frac{\partial}{\partial t} + v_g\frac{\partial}{\partial x_1} - \frac{\partial \Omega}{\partial x_1}\frac{\partial}{\partial k}\right) N = 2\omega_i N \qquad , \tag{IV.2.12}$$

where N is the quasi-particle (or, sometimes called, plasmon) distribution function defined by

$$N(k,x_1,t_1) = p\,\frac{\partial D_r}{\partial \Omega} = \int d\omega\ |\delta\hat{E}|^2\ \frac{\partial D_r}{\partial \omega} \quad , \tag{IV.2.13}$$

and ω_i is the linear damping or growth rate; i.e.,

$$\omega_i = -\frac{D_i}{(\partial D_r/\partial \Omega)} \quad . \tag{IV.2.14}$$

Equation (IV.2.12) is the desired wave kinetic equation including the effect of a small but finite D_i. Thus, with $\omega_i \neq 0$, the number of quasi particles is no longer conserved. Instead, it can be either dissipated or emitted along the ray path. Equation (IV.2.12), therefore, is useful in calculating the wave energy deposition with radio-frequency wave heating.

§IV.3 Cutoff, resonance and WKB connection formulae

[Ref. Stix and Swanson, PPPL-1903, Chapte 10 of Stix's book]

As indicated by Eq. (IV.1.19), the geometric optics or eikonal description is based on the assumption that there exist two scales in both time and space. In many cases of practical interest, this two-scale assumption is not uniformly valid and, in the region it breaks down, one has to employ the wave optics; i.e., the full wave equation. This is the subject of the present section.

Since, in confined plasmas, the plasma properties generally evolve much slowlier than the wave phenomena of interest to us, we shall, for the purpose of this discussion, assume the plasma to be time-stationary but inhomogeneous. The corresponding validity condition for the geometric optics is simply the two space-scale assumpiton, i.e.,

$$\left|\frac{d\underset{\sim}{k}}{d\underset{\sim}{x}}\right| << |\underset{\sim}{k}|^2 \quad . \tag{IV.3.1}$$

Two cases often occurred in plasma physics such that Eq. (IV.3.1) breaks down are cutoffs and resonances where $|\underset{\sim}{k}| \to 0$ and $|\underset{\sim}{k}| \to \infty$, respectively. We shall consider both cases via examples.

As an example for the cutoff, let us consider the following wave equation

$$(\frac{d^2}{dx^2} + \kappa x)\delta\hat{E}(x) = 0 \quad . \tag{IV.3.2}$$

From the geometric optics, we have $|k(x)| = |\kappa x|^{1/2}$ and, thereby, $|k'(x)| = |dk/dx| = 1/2\ |\kappa^{1/2}\ x^{-1/2}|$. Thus, as $|x| \to 0$, we find

$$\lim_{|x|\to 0} |k'| = \frac{1}{2}|\frac{\kappa}{x}|^{1/2} >> |k|^2 = |\kappa x| \quad ; \tag{IV.3.3}$$

i.e,., the two-scale assumption breaks down near $x = 0$. We note that Eq. (IV.3.2) indicates that the wave is propagating for $\kappa x > 0$ and evanescent for $\kappa x < 0$. Thus, wave reflection occurs at $x = 0$ and, in terms of quasi particles, $x = 0$ is called the regular turning point. Just a reminder to the readers, Eq. (IV.3.2) is called the Airy equation and the solutions, Airy functions, are well tabulated.

We now illustrate resonances via the following wave equation

$$(\frac{d^2}{dx^2} + \frac{\mu}{x})\delta\hat{E}(x) = 0 \quad . \tag{IV.3.4}$$

Here, we have $|k(x)| = |\mu/x|^{1/2}$ and $|k'| = |\mu/x^3|^{1/2}/2$. Thus, again, we have, as $|x| \to 0$,

$$\lim_{|x|\to 0} |k'| \propto |x|^{-3/2} \gg |k|^2 \propto |x|^{-1} \quad ; \tag{IV.3.5}$$

that is, the two-scale assumption breaks down near $x = 0$. In this case, $|k| \to \infty$ as $x \to 0$ and $x = 0$ is called the singular turning point. Since, quite often, the solution is singular at $x = 0$, $|\delta\hat{E}|$ tends to become very large at $x = 0$, the resonance point (layer). Thus, in the presence of small but finite dissipaton, we may expect significant wave absorption at $x = 0$ and, thus, resonances play crucial roles in wave heating schemes. We remark that this phenomenon is called resonant absorption and we shall discuss it in more details in Sec. §IV.4. Another feature associated with resonances is that $|k| \to \infty$ near $x = 0$. We, therefore, can also expect microscopic scales, neglected in the physical model, should be included near the resonances. This consideration is the basis of the so-called linear mode conversion process to be discussed in Sec. §IV.5.

In the above discussions, the full wave solutions are only required near cutoffs or resonances; say, at $x = 0$. Away from $x = 0$, the geometric optics (i.e., eikonal or WKB approximations) remains valid. Thus, by asymptotically matching the full wave solutions with the WKB solutions, we can obtain a relation between the WKB solutions valid for $x > 0$ and the WKB solutions valid for $x < 0$. This relation is called the WKB connection formula.

We shall illustrate the derivation of this relation for the following case of a second-order wave equation with a regular turning point at $x = 0$. That is, we have

$$\left(\frac{d^2}{dx^2} + f^2(x)\right)\delta\hat{E}(x) = 0 \quad , \tag{IV.3.6}$$

and

$$f^2(x) \begin{cases} > 0 \text{ for } x > 0 & , \\ = -g^2(x) < 0 \text{ for } x < 0 & , \\ \simeq \kappa x, \ \kappa > 0 \text{ for } x \simeq 0 & . \end{cases} \qquad \text{(IV.3.7)}$$

From Eqs. (IV.3.6) and (IV.3.7), we can write down the WKB solutions to be, for $x > 0$,

$$\delta E_W^> = c_1 f^{-1/2} \exp(i\,\xi_W^>) + c_2 f^{-1/2} \exp(-i\xi_W^>) \quad ; \qquad \text{(IV.3.8)}$$

and, for $x < 0$,

$$\delta E_W^< = D_1 g^{-1/2} \exp(\xi_W^<) + D_2 g^{-1/2} \exp(-\xi_W^<) \quad ; \qquad \text{(IV.3.9)}$$

where $\xi_W^> = \int_0^x f\,dx$ and $\xi_W^< = \int_0^x g\,dx$.

Near $x = 0$, Eq. (IV.3.6) reduces to the Airy equation with the solutions given by, for $x > 0$,

$$\delta E_f^> = A_1 x^{1/2} J_{1/3}(\xi_1) + B_1 x^{1/2} J_{-1/3}(\xi_1) \quad , \qquad \text{(IV.3.10)}$$

and, for $x < 0$,

$$\delta E_f^< = A_2 |x|^{1/2} I_{1/3}(\xi_2) + B_2 |x|^{1/2} I_{-1/3}(\xi_2) \quad ; \qquad \text{(IV.3.11)}$$

where

$$\xi_1 = \frac{2}{3} \kappa^{1/2} x^{3/2} \quad , \tag{IV.3.12}$$

and

$$\xi_2 = \frac{2}{3} \kappa^{1/2} |x|^{3/2} \quad . \tag{IV.3.13}$$

Now at $x = 0$, we impose the continuity conditions, $\delta E_f^{>} = \delta E_f^{<}$ and $d\delta E_f^{>}/dx = d\delta E_f^{<}/dx$, to obtain

$$A_2 = -A_1 \text{ and } B_2 = B_1 \quad . \tag{IV.3.14}$$

We now perform asymptotic matching between δE_f and δE_W. For $x > 0$, we have as $x \to \infty$

$$\delta E_f^{>} \to \left(\frac{2}{\pi}\right)^{1/2} \left(\frac{x}{\xi_1}\right)^{1/2} \left[A_1 \cos\left(\xi_1 - \frac{5\pi}{12}\right) + B_1 \cos\left(\xi_1 - \frac{\pi}{12}\right)\right] \quad . \tag{IV.3.15}$$

Meanwhile, noting that

$$\lim_{x \to 0} \xi_W^{>} \to \xi_1 \tag{IV.3.16}$$

and

$$\lim_{x \to 0} f^{-1/2} \to \left(\frac{2x}{3\xi_1}\right)^{1/2} \quad ; \tag{IV.3.17}$$

we have, as $x \to 0$,

$$\delta E_W^{>} \rightarrow \left(\frac{2}{3}\right)^{1/2}\left(\frac{x}{\xi_1}\right)^{1/2}\left[c_1 \exp(i\xi_1) + c_2 \exp(-i\xi_1)\right] \quad . \qquad \text{(IV.3.18)}$$

Equating Eqs. (IV.3.15) and (IV.3.18), we find

$$c_1 = A \exp\left(- i \frac{5\pi}{12}\right) + B \exp\left(- i \frac{\pi}{12}\right) \quad , \qquad \text{(IV.3.19)}$$

and

$$c_2 = A \exp\left(\frac{i5\pi}{12}\right) + B \exp\left(\frac{i\pi}{12}\right) \quad . \qquad \text{(IV.3.20)}$$

Here, $A \equiv \frac{1}{2}\,(3/\pi)^{1/2} A_1$ and $B \equiv \frac{1}{2}\,(3/\pi)^{1/2} B_1$.

We can carry out a similar analysis for $x < 0$ and obtain, omitting the details,

$$D_1 = - A + B \quad , \qquad \text{(IV.3.21)}$$

and

$$D_2 = [- A \exp(-i5\pi/6) + B \exp(-i\pi/6)] \quad . \qquad \text{(IV.3.22)}$$

Combining the above results, we arrive at the following connection formula for WKB solutions across a regular turning point

$$f^{-1/2}\{[A \exp(- \frac{i5\pi}{12}) + B \exp(\frac{-i\pi}{12})]\exp(i\xi_W^{>}) + [A \exp(\frac{i5\pi}{12})$$

$$+ B \exp(\frac{i\pi}{12})]\exp(-i\xi_W^>)\}$$

$$\leftrightarrow g^{-1/2}\{(-A + B)\exp(\xi_W^<) + [-A \exp(\frac{-i5\pi}{6}) + B \exp(\frac{-i\pi}{6})] \exp(-\xi_W^<)\} \quad . \tag{IV.3.23}$$

We note that, using the same technique, we can also derive a WKB connection formula across a singular turning point. This is done in Stix's book and we refer the readers to it.

We now shall divert somewhat and make two remarks. First, we note that the connection formula given by Eq. (IV.3.23) can be used to derive the Bohm-Sommerfeld quantization condition, i.e.,

$$\int_{T_1}^{T_2} f(z)dz = (n + \frac{1}{2})\pi \qquad \text{for } n = 0, 1, \ldots \quad . \tag{IV.3.24}$$

Here, $z = x + iy$; i.e., we have generalized the wave equation, Eq. (IV.3.6), to the complex plane. T_1 and T_2 are two turning points; that is, $f(T_1) = f(T_2) = 0$. Again, T_1 and T_2 are generally complex. Referring to Fig. (IV.3.1),

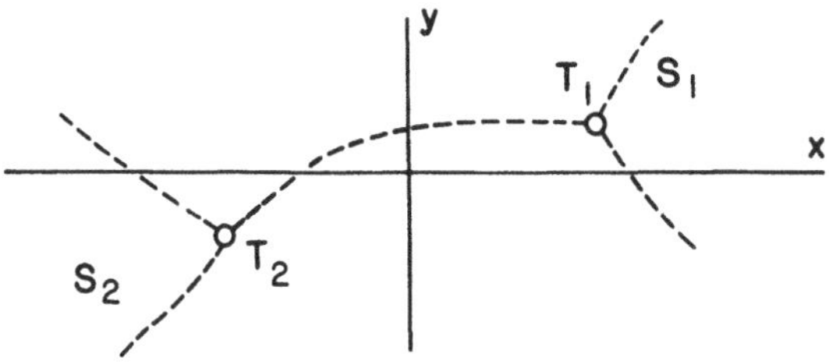

Fig. (IV.3.1) The Stokes diagram in the complex (x,y) plane.

we see that Eq. (IV.3.24) is applicable to solutions which are subdominant in regions S_1 and S_2. That is, solutions in S_1 and S_2 decay away from T_1 and T_2. Meanwhile, the dashed lines in Fig. (IV.3.1) corresponds to the anti-Stokes lines, i.e., the WKB solutions are purely oscillatory along the lines. For a detailed analysis of WKB technique in the complex plane, we refer the readers to the monograph "An Introduction to Phase-Integral Method," by J. Heading, John Wiley, and Sons, Inc., N.Y. (1962). We note that the WKB technique is useful in solving eigenvalue problems.

The second remark is regarding the derivation of asymptotic behaviors of the Airy functions (hence, the WKB connection formula) via the Laplace method. The main purpose is to introduce the Laplace method which is often useful in dealing problems involving asymptotic matchings. Recalling the Airy equation to be

$$\left(\frac{d^2}{dx^2} + \kappa x\right)\delta\hat{E}(x) = 0 \quad , \qquad \text{(IV.3.2)}$$

we then define the Laplace transform to be

$$\delta\hat{E}(x) = \int_{Q_1}^{Q_2} d\omega\; p(\omega)\, \exp(i\omega x) \quad . \qquad \text{(IV.3.25)}$$

Equation (IV.3.25) into Eq. (IV.3.2), we readily derive

$$p(\omega) = A \exp\left(-\frac{i\omega^3}{3}\right) \quad , \qquad \text{(IV.3.26)}$$

if

$$p(\omega)\exp(i\omega x)\Big|_{Q_1}^{Q_2} = 0 \quad . \qquad \text{(IV.3.27)}$$

By examining the $|\omega| \to \infty$ limit of $p(\omega)$, it is easy to show that Q_1 and Q_2 must lie inside the cross-hatched regions shown in Fig. (IV.3.2).

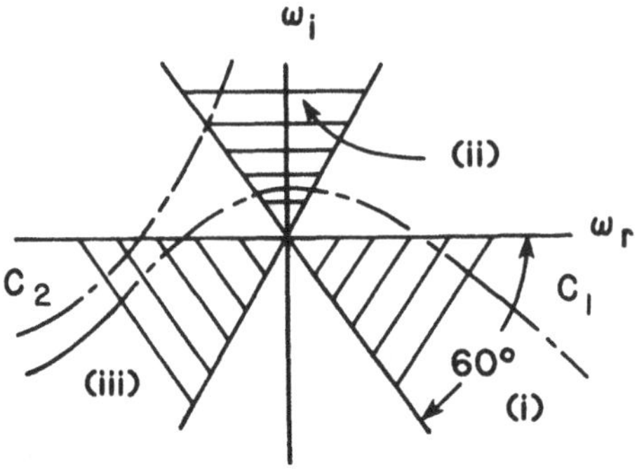

Fig. (IV.3.2) Integration contours for the Laplace integral of Eq. (IV.3.25).

By the proper choice of Q_1 and Q_2, one finds that there exist <u>two</u> independent integration contours denoted as C_1 and C_2 in Fig. (IV.3.2). Thus, C_1 corresponds to Q_1 in (iii) and Q_2 in (i). Meanwhile, for C_2, we have Q_1 in (iii) and Q_2 in (ii). This should be expected because Eq. (IV.3.2) is a second-order differential equation with, therefore, two independent solutions. With the integration path specified, we have

$$\delta\hat{E}(x) = \int_{C_{1,2}} d\omega \, \exp\left[ix\left(\omega - \frac{\omega^3}{3x}\right)\right] \quad . \tag{IV.3.28}$$

With $\delta\hat{E}(x)$ given by an integral representation, the $|x| \to \infty$ asymptotic behaviors can then be obtained using the saddle-point metho (i.e., the method of steepest descents).

Recapitulating the result of the saddle-point method, we have

$$\lim_{|\alpha|\to\infty} I(\alpha) = \lim_{|\alpha|\to\infty} \int_C \exp[\alpha\beta(z)]dz$$

$$\to \left(\frac{2\pi}{\alpha\rho}\right)^{1/2} \exp\left(i\phi + \alpha\beta(z_o)\right) \quad , \tag{IV.3.29}$$

where z_0 is the saddle point given by

$$\beta'(z_0) = 0 \quad , \tag{IV.3.30}$$

$$\beta''(z_0) \equiv \rho \exp(i\theta) \tag{IV.3.31}$$

and

$$\phi = \pm \frac{\pi}{2} - \frac{\theta}{2} \tag{IV.3.32}$$

is the inclination angle of the path with respect to the saddle point. In Eq. (IV.3.32) we should choose the proper ϕ such that the path corresponds to the steepest descent, i.e., from valley to valley, instead from ridge to ridge.

We now illustrate the derivation of the asymptotic behaviors of $\delta\hat{E}(x)$ by adopting the C_1 integration contour. For $x > 0$, we have, as $x >> 1$, two saddle points at $x^{1/2}$ and $-x^{1/2}$ with $\phi = -\pi/4$ and $\pi/4$, respectively. Applying Eq. (IV.3.29), we thus, derive

$$\begin{aligned}\delta E^{>} &\to \left(\frac{2\pi x^{1/2}}{2x}\right)^{1/2}\{\exp(\tfrac{2}{3}\, ix^{3/2} - i\,\tfrac{\pi}{4}) + \exp[-(\tfrac{2}{3}\, ix^{3/2} - i\,\tfrac{\pi}{4})]\} \\ &= \left(\frac{2\pi^{1/2}}{x^{1/4}}\right) \cos(\tfrac{2}{3}\, x^{3/2} - \tfrac{\pi}{4}) \quad .\end{aligned} \tag{IV.3.33}$$

Meanwhile, for $x < 0$ and $|x| >> 1$, we have the saddle points at $\pm\, i|x|^{1/2}$. For the C_1 contour, however, only the saddle point at $-i|x|^{1/2}$ is relevant and, correspondingly, $\phi = 0$. We, thus, derive

$$\delta E^{<} \rightarrow \pi^{1/2} |x|^{-1/4} \exp(-\frac{2}{3} |x|^{3/2}) \quad . \qquad \text{(IV.3.34)}$$

We note that the connection formula given by Eqs. (IV.3.33) and (IV.3.34) agrees with that given by Eq. (IV.3.23) for the case A = B. The case with $A \neq B$ can be derived by adopting the C_2 integration contour.

§IV.4 Tonk-Dattner resonance and resonant absorption

In this section, we shall illustrate some physics associated with resonances via the example of the Tonk-Dattner resonance. In this model, we consider a cold, unmagnetized, inhomogeneous plasmas located between two capacitor plates which are driven externally with the following electric field

$$\delta \underset{\sim}{E}_0 = \delta \hat{\underset{\sim}{E}}o \exp(-i\omega_0 t) + c.c. \qquad \text{(IV.4.1)}$$

Here, $|\omega_0|^2 >> |\omega_{pi,max}|^2$ such that ion dynamics is negligible. Furthermore, $|\delta \underset{\sim}{E}_0|$ is taken to be sufficiently small to justify a linear description.

Within this model, the electron dynamics is, thus, describable by the Poisson's equation and the linearized equations of continuity and motion without the thermal pressure term; that is, suppressing the index for electrons,

$$\frac{\partial}{\partial \underset{\sim}{x}} \cdot \delta \underset{\sim}{E} = -4\pi e \, \delta n \quad , \qquad \text{(IV.4.2)}$$

$$\frac{\partial}{\partial t} \delta n + \frac{\partial}{\partial \underset{\sim}{x}} \cdot (n_0 \delta \underset{\sim}{v}) = 0 \quad , \qquad \text{(IV.4.3)}$$

and

$$\frac{\partial}{\partial t}\,\delta \underset{\sim}{v} = -\,\frac{e}{m}\,\delta \underset{\sim}{E} \qquad . \qquad \text{(IV.4.4)}$$

Here, $n_o = n_o(\underset{\sim}{x})$. Noting that Eqs. (IV.4.3) and (IV.4.4) can be combined into

$$\frac{\partial^2 \delta n}{\partial t^2} - \frac{e}{m}\,\frac{\partial}{\partial \underset{\sim}{x}} \cdot (n_o \delta \underset{\sim}{E}) = 0 \qquad . \qquad \text{(IV.4.5)}$$

Using Eq. (IV.4.2), Eq. (IV.4.5) can be further reduced to the following wave equation

$$\frac{\partial}{\partial \underset{\sim}{x}} \cdot \left(\frac{\partial^2}{\partial t^2} + \omega_{pe}^2(\underset{\sim}{x})\right) \cdot \delta \underset{\sim}{E}(\underset{\sim}{x}) = 0 \qquad . \qquad \text{(IV.4.6)}$$

Expressing $\delta \underset{\sim}{E}$ as

$$\delta \underset{\sim}{E}(\underset{\sim}{x}) = \delta \hat{\underset{\sim}{E}}(\underset{\sim}{x}) \exp(-i\omega_o t) + \text{c.c.} \qquad , \qquad \text{(IV.4.7)}$$

Eq. (IV.4.6) yields

$$\frac{\partial}{\partial \underset{\sim}{x}} \cdot D(\underset{\sim}{x})\, \delta \hat{\underset{\sim}{E}}(\underset{\sim}{x}) = 0 \qquad , \qquad \text{(IV.4.8)}$$

where

$$D(\underset{\sim}{x}) = 1 - \frac{\omega_{pe}^2(\underset{\sim}{x})}{\omega_o^2} \qquad . \qquad \text{(IV.4.9)}$$

For simplicity and without any loss of physics, let us consider the one-dimensional limit; i.e., $\delta \hat{\underset{\sim}{E}} = \delta \hat{E} \underset{\sim}{e}_x$, $n_o = n_o(x)$ and etc. Correspondingly, Eq. (IV.4.8) becomes

$$\frac{d}{dx}\left(D(x)\delta\hat{E}\right) = 0 \quad , \tag{IV.4.10}$$

or, equivalently,

$$\frac{d\delta\hat{E}}{dx} + \frac{D'}{D}\,\delta\hat{E} = 0 \quad . \tag{IV.4.11}$$

Here $D' \equiv dD(x)/dx$. From Eq. (IV.4.11), we see that the WKB wavenumber is given by

$$k(x) = -\frac{iD'}{D} \quad . \tag{IV.4.12}$$

Thus, if, at $x = x_0$, we have

$$D(x_0) = 1 - \frac{\omega_{pe}^2(x_0)}{\omega_0^2} = 0 \quad , \tag{IV.4.13}$$

<u>and</u>

$$D'(x_0) \neq 0 \quad ; \tag{IV.4.14}$$

then

$$\lim_{x \to x_0} |k(x)| \to \infty \quad , \tag{IV.4.15}$$

and we have a resonance at $x = x_0$. In other words, the Tonk-Dattner resonances occur when the external driving frequency equals to the local electron plasma frequency. Equation (IV.4.10), of course, can be solved directly and we find

$$\delta\hat{E}(x) = \frac{\delta\hat{E}_o}{D(x)} \quad . \tag{IV.4.16}$$

In obtaining Eq. (IV.4.16), we have imposed the condition that $\delta\hat{E}(x) = \delta\hat{E}_o$ for $D(x) = 1$; i.e., in the vacuum. Since $D(x_o) = 0$, Eq. (IV.4.16) clearly shows that $\delta\hat{E}(x)$ is singular at $x = x_o$ and, as indicated in Sec. §IV.3, we can expect significant wave absorption near $x = x_o$.

To calculate the resonant absorption, we note that from the Poynting's conservation theorem, Eq. (I.2.9), it is easy to show

$$\frac{d\langle p_x\rangle}{dx} = -\left(\frac{\omega_o}{8\pi}\right)\mathrm{Im}\left[\delta\hat{E}^*\left(\delta\hat{E} + \frac{i4\pi\delta\hat{J}}{\omega_o}\right)\right] \quad , \tag{IV.4.17}$$

where

$$\langle p_x\rangle = \left(\frac{c}{4\pi}\right)\,\mathrm{Re}\left(\delta\hat{\underset{\sim}{E}} \times \delta\hat{\underset{\sim}{B}}^*\right)_x \quad . \tag{IV.4.18}$$

From the equation of motion, Eq. (IV.4.4), we have

$$\delta\hat{J} = \frac{i\omega_{pe}^2(x)\;\delta\hat{E}}{\omega_o} \quad . \tag{IV.4.19}$$

Equation (IV.4.17) then becomes

$$\frac{d\langle p_x\rangle}{dx} = -\left(\frac{\omega_o}{8\pi}\right)\mathrm{Im}\left(\delta\hat{E}^* D(x)\delta\hat{E}\right) \tag{IV.4.20}$$

or, noting Eq. (IV.4.16),

$$\frac{d\langle p_x\rangle}{dx} = -\left(\frac{\omega_o}{8\pi}\right)|\delta\hat{E}_o|^2 \, \mathrm{Im}\left(\frac{1}{D^*(x)}\right) \quad . \tag{IV.4.21}$$

To go further, let us assume the dielectric constant contains a small but finite anti-Hermitian component due to, e.g., dissipations. Thus, replacing ω_o^2 by $\omega_o(\omega_o + i\nu)$ with $|\nu/\omega_o| << 1$, we have

$$D(x) \to 1 - \frac{\omega_{pe}^2}{\omega_o(\omega_o+i\nu)} \simeq 1- \frac{\omega_{pe}^2}{\omega_o^2} + \frac{i(\nu/\omega_o)\omega_{pe}^2(x)}{\omega_o^2} \quad . \tag{IV.4.22}$$

Furthermore, from Eq. (IV.4.21), we note that $d\langle p_x\rangle/dx$ peaks about $x = x_o$ where $\omega_o^2 = \omega_{pe}^2(x_o)$. Since $\mathrm{Re}\, D'(x_o) \neq 0$, we, thus, have for $x \simeq x_o$

$$D(x) \simeq - \left[\left(\frac{2}{\omega_o}\right)\omega_{pe}'(x_o)\right](x - x_o) + \frac{i\nu}{\omega_o} \equiv \kappa(x - x_o + i\eta) \quad . \tag{IV.4.23}$$

Here, for definitiveness, we take $\kappa > 0$ and $\eta = \nu/(\omega_o\kappa)$. Substituting Eq. (IV.4.23) into Eq. (IV.4.21), we find

$$\lim_{|\nu|\to 0} \frac{d\langle p_x\rangle}{dx} = \lim_{|\eta|\to 0} \left(- \frac{\omega_o}{8\pi}\right) \frac{|\delta\hat{E}_o|^2}{\kappa} \, \mathrm{Im}\left(\frac{1}{x-x_o-i\eta}\right) \quad . \tag{IV.4.24}$$

Noting that

$$\lim_{|\eta|\to 0} \frac{1}{x-x_o-i\eta} = P\left(\frac{1}{x-x_o}\right) + i\pi\delta(x - x_o)\mathrm{sgn}(\eta)$$

$$= P\left(\frac{1}{x-x_o}\right) + i\pi\delta(x - x_o)\mathrm{sgn}(\nu) \quad , \tag{IV.4.25}$$

Eq. (IV.4.24) becomes

$$\lim_{|\nu|\to 0} \frac{d\langle p_x\rangle}{dx} = - \frac{\omega_o}{8\pi} \frac{|\delta\hat{E}_o|^2}{\kappa} \pi\delta(x - x_o)\mathrm{sgn}(\nu) \quad . \tag{IV.4.26}$$

For positive dissipations, $\nu > 0$, and $n_o(x)$ sketched in Fig. (IV.4.1a), the Poynting flux moving in the negative-x direction then suffers a loss at $x = x_o$ given by

$$\Delta P = \frac{\omega_o|\hat{E}_o|^2}{(8\kappa)} \quad , \qquad \text{(IV.4.27)}$$

which is the rate of energy resonantly absorbed [referring to Fig. (IV.4.1b)].

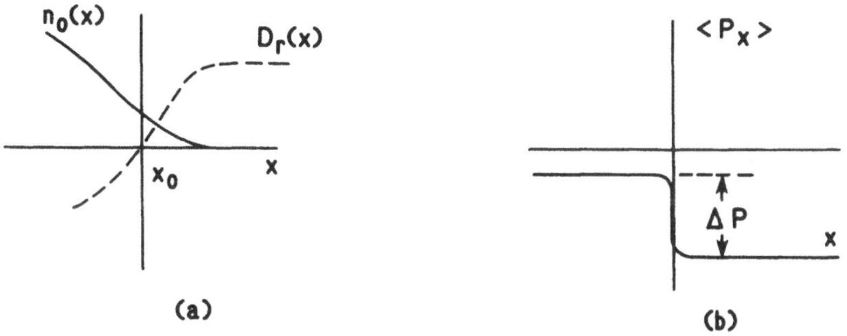

Fig. (IV.4.1) Sketches of (a) density n_o, Real (D) of Eq. (IV.4.22), and (b) Poynting flux vs. x in Tonk-Dattner resonance. x_o is the resonance point.

§IV.5 Linear mode conversion

As indicated in Secs. §IV.3 and, specifically, §IV.4 for the Tonk-Dattner resonance, wave perturbations become singular and $|\underset{\sim}{k}| \to \infty$ near resonances. Thus, dynamics with microscopic space scales neglected in the original physical model should, in principle, be included near the resonances and this can lead to the so-called linear mode-conversion process. Here, we shall illustrate this process by pursuing the model of Tonk-Dattner resonances.

In this model where the plasmas are unmagnetized and ions are assumed to be immobile, the relevant microscopic space scale is the Debye length, λ_D, which enters the dynamics via the thermal pressure term in the electron equation of motion and the equation of state, i.e.,

$$\frac{\partial}{\partial t}\,\delta \underset{\sim}{v} = -\frac{e}{m}\,\delta \underset{\sim}{E} - \frac{1}{mn_o}\,\frac{\partial}{\partial \underset{\sim}{x}}\,\delta p \quad , \tag{IV.5.1}$$

and

$$\delta p = \frac{\gamma p_o \delta n}{n_o} \quad . \tag{IV.5.2}$$

Here, γ is the ratio of specific heat. Combining Eqs. (IV.5.1) and (IV.5.2) with the equation of continuity and the Poisson's equation, i.e., Eqs. (IV.4.3) and (IV.4.2), we can readily derive the following wave equation

$$\underset{\sim}{\nabla} \cdot \left(\bar{\lambda}_D^{-2}\nabla^2 + D(\underset{\sim}{x})\right)\delta\hat{\underset{\sim}{E}}(\underset{\sim}{x}) = 0 \quad ; \tag{IV.5.3}$$

here, $\bar{\lambda}_D^{-2} = v_{te}^2/\omega_o^2$, $D(\underset{\sim}{x})$ is given by Eq. (IV.4.9), and $\underset{\sim}{\nabla} T_e = 0$ is assumed. Note that Eq. (IV.5.3) reduces to Eq. (IV.4.8) in the cold plasma ($v_{te} \to 0$) limit.

Again, the physics can be best elucidated in the one-dimensional limit. In this limit, Eq. (IV.5.3) becomes

$$\left(\lambda_D^{-2}\,\frac{d^2}{dx^2} + D(x)\right)\delta\hat{E}(x) = \delta\hat{E}_o \quad . \tag{IV.5.4}$$

Equation (IV.5.4) immediately contrasts the cold plasma ($\bar{\lambda}_D = 0$) limit [i.e., Eq. (IV.4.16)] in that $\delta\hat{E}(x)$ is, in general, no longer singular at $x = x_o$ where $D(x_o) = 0$. Furthermore, noting that Eq. (IV.5.4) is a second-order inhomogeneous differential equation, we shall make the following observations.

(i) The two homogeneous solutions of Eq. (IV.5.4) correspond to Bohm-Gross waves which are propagating (evanescent or divergent) for $D(x) > 0$ (<0). (ii) For $|x - x_0| >> \bar{\lambda}_D$, the Bohm-Gross waves are decoupled from the particular solution approximately given by $\delta\hat{E}_p(x) \simeq \delta\hat{E}_0/D(x)$. Alternatively, we may call the Bohm-Gross waves the warm plasma (or short wavelength) mode. (iii) For $|x - x_o| \sim O(\bar{\lambda}_D)$, the warm plasma modes and the cold plasma mode are strongly coupled. This coupling occurs since as $x \to x_0$, we have $|k_{warm}| \to 0$ and $|k_{cold}| \to \infty$ and we expect the two k's of the same order for x near x_0.

To study the effect of coupling in more details, we let, for $x \simeq x_0$, $D(x) \simeq \kappa(x - x_0) \equiv \kappa t$ and $|\bar{\lambda}_D\kappa| << 1$. Equation (IV.5.4) then becomes

$$\left(\bar{\lambda}_D^2 \frac{d^2}{dt^2} + \kappa t\right) \delta\hat{E} = \delta\hat{E}_0 \quad . \tag{IV.5.5}$$

Equation (IV.5.5), thus, corresponds to the inhomogeneous Airy equation. The solutions are tabulated [c.f. Ch. 10 of Abramowitz and Stegun] and are given by

$$\delta\hat{E}(x) = - \frac{\pi\delta\hat{E}_0}{(\kappa\bar{\lambda}_D)^{2/3}} \left[c_1 A_i\left(- \frac{t}{\Delta}\right) + c_2 B_i\left(- \frac{t}{\Delta}\right) + G_i\left(- \frac{t}{\Delta}\right)\right] . \tag{IV.5.6}$$

Here, $\Delta \equiv \left(\bar{\lambda}_D^2/\kappa\right)^{1/3} >> \bar{\lambda}_D$ is, sometimes called, the Airy scale length.

To determine the two constants, c_1 and c_2, we need to impose two boundary conditions. Referring to Fig. (IV.4.1a), one condition is clear that for $x < x_0$, there should be no short wavelength exponentially divergent solution. The other condition concerns with the two Bohm-Gross waves propagating in the $x > x_0$ region with $v_g > 0$ and $v_g < 0$, respectively. Since there is no short wavelength external source at $x > x_0$, it rules out the possibility of external excitation of the $v_g < 0$ Bohm-Gross wave. On the other hand, we observe that, as $n_0(x) \to 0$, $|k(x)\bar{\lambda}_D| \to O(1)$ and the Bohm-Gross waves suffer strong electron Landau damping in the low-density region. Thus, as the $v_g > 0$ Bohm-Gross wave propagates from the resonance region at $x = x_0$ outward to the low-density region, it will be strongly damped and give rise to negligible reflection. That is, the other boundary condition corresponds to the perfect absorption of the $v_g > 0$ Bohm-Gross wave; i.e., there exists no Bohm-Gross wave with $v_g < 0$ in the $x > x_0$ region. With these two boundary conditions specified, it is then straightforward to show that

$$c_1 = i \quad \text{and} \quad c_2 = 0 \quad . \tag{IV.5.7}$$

The corresponding asymptotic behaviors are then given by the following expressions

$$\delta E^{>} \to - \frac{\pi \delta\hat{E}_0}{(\kappa\bar{\lambda}_D)^{2/3}} \left(\frac{\Delta}{x-x_0}\right)^{1/4} \exp\{i[\tfrac{2}{3}(\tfrac{x-x_0}{\Delta})^{3/2} - \tfrac{\pi}{4}]\} + \frac{\delta\hat{E}_0}{\kappa(x-x_0)} \tag{IV.5.8}$$

for $x > x_0$, and

$$\delta E^{<} \to \frac{\delta\hat{E}_0}{\kappa(x-x_0)} \tag{IV.5.9}$$

for $x < x_0$. Equations (IV.5.6), (IV.5.8) and (IV.5.9) indicate that introducing the microscopic dynamics via finite $\bar{\lambda}_D$ removes the singularity at $x = x_0$ even without dissipation. The field amplitude for $|x - x_0| \sim O(\Delta)$ is still significantly amplified over $\delta\hat{E}_0$, i.e., $|\delta E|/|\delta\hat{E}_o| \sim O(\kappa\bar{\lambda}_D|^{-2/3}) >> 1$. Sometimes, this is called Airy amplification (swelling). Furthermore, comparing to the cold plasma result, Eq. (IV.5.8) contains an additional term corresponding to a warm plasma Bohm-Gross wave propagating away from the resonance (layer) toward the low-density region. This process of exciting short wavelength modes due to the resonance of long wavelength modes is called linear mode conversion (or transformation in certain literatures). Now, as long as the wave energy carried by the $v_g > 0$ Bohm-Gross wave is perfectly absorbed via electron Landau damping, we remark that the rate of resonant energy absorption is the same as that given by the cold plasma theory, i.e., Eq. (IV.4.27).

We now briefly discuss another example which concerns the linear mode conversion of ion Bernstein wave [Ref. Ono, PPPL-1593 (1979), Ono and Wong, PPPL-1670 (1980)]. For this case, $\underset{\sim}{B}_o = B_o \underset{\sim}{e}_z$ and $n_o = n_o(x)$ [c.f. Fig. IV.5.1)].

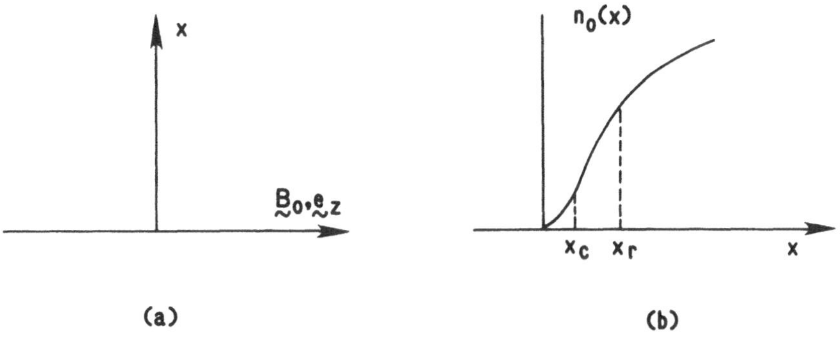

Fig. (IV.5.1) Sketches of (a) magnetic field $\underset{\sim}{B}_o$ and nonuniformity direction, x, and (b) density n_o profile vs. x for the linear mode conversion of ion Bernstein waves.

Meanwhile, the external driving frequency is of the order of ion cyclotron frequency; typically, $\omega_o \lesssim 2\Omega_i$. Let us first consider the wave propagation characteristics using the cold plasma description. Here, the corresponding wave equation is the following

$$\left(\frac{\partial}{\partial x} D_{\perp c} \frac{\partial}{\partial x} + \frac{\partial}{\partial z} D_{\parallel} \frac{\partial}{\partial z}\right)\delta\phi(x,z) = 0 \quad , \tag{IV.5.10}$$

where

$$D_{\perp c}(x) = 1 + \frac{\omega_{pe}^2(x)}{\Omega_e^2} - \frac{\omega_{pi}^2(x)}{\omega_o^2-\Omega_i^2} \quad , \tag{IV.5.11}$$

$$D_{\parallel}(x) = 1 - \frac{\omega_{pe}^2(x)}{\omega_o^2} \quad , \tag{IV.5.12}$$

and the electrostatic approximation has been made. In the eikonal representation, Eq. (IV.5.10) then yields

$$k^2(x) = - k_{\parallel}^2 \frac{D_{\parallel}}{D_{\perp c}} \quad . \tag{IV.5.13}$$

For very low densities, we have $\omega_{pe}^2(x) << \Omega_e^2$ and $\omega_{pi}^2(x) << \Omega_i^2$, i.e., $D_{\perp c} \simeq 1$ and, hence,

$$k^2(x) \simeq - k_{\parallel}^2 D_{\parallel}(x) \quad . \tag{IV.5.14}$$

Noting Eq. (IV.5.12), Eq. (IV.5.14), thus, predicts the existence of a cutoff at $x = x_c$ where $\omega_o^2 = \omega_{pe}^2(x_c)$. Since, for $x < x_c$, $D_{\parallel} \simeq 1$ and $k^2(x) \simeq -k_{\parallel}^2$, the wave has to tunnel into the plasma. This tunneling, however, has little effect because, typically $|k_{\parallel} x_c| << 1$. As we move beyond x_c such that $D_{\parallel}(x) \simeq -\omega_{pe}^2(x)/\omega_o^2$, we have $k^2 \simeq \omega_{pe}^2(x) k_{\parallel}^2/\omega_o^2$; i.e., the wave has the characteristics of the magnetized electron plasma (Gould-Trievelpiece) wave. Moving further into the plasma, we have $\omega_{pi} \sim \omega_o$ and, hence, the possibility of a resonance at $x = x_r$ where $D_{\perp c}(x_r) = 0$. Noting that, typically, this resonance is located near the plasma edge and $\omega_o \lesssim 2\Omega_i$, the resonance condition can be approximated as

$$\omega_{pi}^2(x_r) \simeq \omega_o^2 - \Omega_i^2 \simeq 3\Omega_i^2 \quad . \tag{IV.5.15}$$

We now consider the associated mode-conversion process. Since $\omega_o \sim O(\Omega_i)$ and $\rho_i >> \rho_e$, ion Larmor radius is the relevant microscopic scale. Anticipating the Airy scale length of the order $\rho_i(\kappa\rho_i)^{-1/3} >> \rho_i$, we may take the small ion Larmor-radius expansion. Within the eikonal approximation, we find

$$D_\perp \simeq D_{\perp c} + \frac{(k\rho_i)^2\omega_{pi}^2}{(4\Omega_i^2-\omega_o^2)} \quad , \tag{IV.5.16}$$

and the corresponding WKB wave equation, Eq. (IV.5.10), becomes

$$\frac{\rho_i^2\omega_{pi}^2}{4\Omega_i^2-\omega_o^2} k^4 + D_{\perp c}k^2 + k_\parallel^2 D_\parallel = 0 \quad . \tag{IV.5.17}$$

Noting that $D_{\perp c} > 0$ for $x > x_r$ (i.e., higher density region), Eq. (IV.5.17), thus, shows that, for $\omega_o^2 < 4\Omega_i^2$, the mode-converted short wavelength ion Bernstein wave will propagate into the high-density region. This has the desirable feature that the wave energy will be cyclotron damped by ions in the bulk plasma. Figure (IV.5.2) is a sketch of the WKB wavenumber k^2 vs. x.

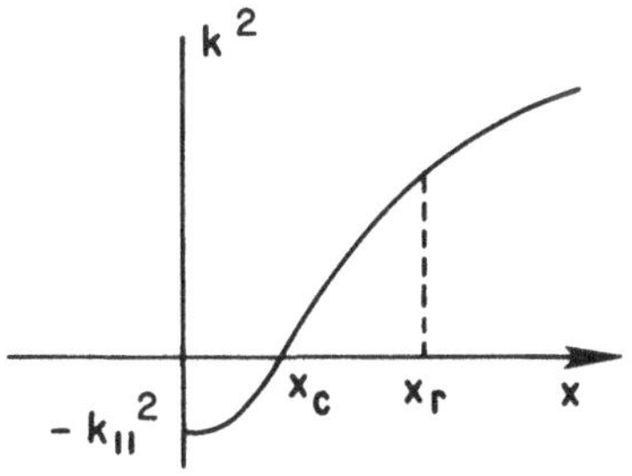

Fig. (IV.5.2) Sketch of WKB wavenumber k^2 vs. x for the linear mode conversion of ion Bernstein waves.

Homework # 7

(1) Use the connection formula for the regular turning point to verify the Bohr-Sommerfeld quantization condition

$$\int_{T_1}^{T_2} f(z)dz = \left(n + \frac{1}{2}\right)\pi \quad ,$$

where the wave equation is $[d^2/dz^2 + f^2(z)]\delta\phi(z) = 0$ and T_1 and T_2 are the turning points.

(2) (2.a) Show that for the Weber equation

$$\left(\frac{d^2}{dz^2} + E - z^2\right)\delta\phi(z) = 0 \quad ,$$

the eigenvalue derived from the WKB (B.-S.) quantization rule is exact.

(2.b) Try to draw the anti-Stokes lines.

3. Use the Laplace method and re-derive the connection formula for the regular turning point such that the solution in the evanescent region is divergent; i.e. taking the C_2 integration path as noted in lectures.

§IV.6 Linear gyrokinetic equation in nonuniform plasmas

[Ref. Rutherford and Frieman, Phys. Fluids 11, 569 (1968); Taylor and Hastie, Plasma Phys. 10, 479 (1968); Antonsen and Lane, Phys. Fluids 23, 1205 (1980); Catto, Plasma Phys. (1978)]

In this section, we shall generalize the derivation of the linear gyrokinetic equation in uniform plasmas (§III.7) to nonuniform plasmas. The crucial assumption is that the nonuniformities are sufficiently weak such that $\rho/L_o \sim O(\eta) << 1$ with η being the smallness parameter. In this respect, η may also be regarded as the adiabaticity parameter. With $\eta << 1$, it can then be expected that, to the lowest order, the unperturbed particle orbit corresponds to that of a guiding center. Thus, we may adopt the same guiding-center transformation (§III.1), which we recall to be

$$\underset{\sim}{X} = \underset{\sim}{x} + \underset{\sim}{v} \times \frac{\underset{\sim}{e}_{\parallel}}{\Omega} \quad , \tag{III.1.6}$$

and

$$\underset{\sim}{V} = (\varepsilon, \mu, \alpha, \sigma) \tag{III.1.7}$$

The complication, of course, enters through the spatial dependence of $\underset{\sim}{B}_o$ and, hence, $\underset{\sim}{e}_{\parallel}$, μ, α, etc. The corresponding transformation relations are then given by

$$\frac{\partial}{\partial \underset{\sim}{x}} \to \frac{\partial}{\partial \underset{\sim}{X}} + \underset{\sim}{\lambda}_{B1} + \underset{\sim}{\lambda}_{B2} \quad , \tag{IV.6.1}$$

and

$$\frac{\partial}{\partial \underset{\sim}{v}} \to \frac{\partial}{\partial \underset{\sim}{V}} + \frac{\underset{\approx}{I} \times \underset{\sim}{e}_{\parallel}}{\Omega} \cdot \frac{\partial}{\partial \underset{\sim}{X}} \quad ; \tag{III.1.11}$$

where

$$\underset{\sim}{\lambda}_{B1} = \underset{\sim}{v} \times \frac{\partial}{\partial \underset{\sim}{x}}\left(\frac{\underset{\sim}{e}_{\parallel}}{\Omega}\right) \cdot \frac{\partial}{\partial \underset{\sim}{X}} \quad , \tag{IV.6.2}$$

$$\boldsymbol{\lambda}_{B2} = \left(\frac{\partial \mu}{\partial \mathbf{x}}\right) \frac{\partial}{\partial \mu} + \left(\frac{\partial \alpha}{\partial \mathbf{x}}\right) \frac{\partial}{\partial \alpha} \quad , \tag{IV.6.3}$$

$$\frac{\partial}{\partial \mathbf{V}} = \mathbf{v} \frac{\partial}{\partial \varepsilon} + \mathbf{v}_{\perp} \frac{\partial}{B_o \partial \mu} + \mathbf{e}_{\alpha} \frac{1}{v_{\perp}} \frac{\partial}{\partial \alpha} \quad , \tag{III.1.12}$$

$$\frac{\partial \mu}{\partial \mathbf{x}} = \frac{-[\mu(\partial B_o/\partial \mathbf{x}) + v_{\parallel}(\partial \mathbf{e}_{\parallel}/\partial \mathbf{x}) \cdot \mathbf{v}_{\perp}]}{B_o} \quad , \tag{IV.6.4}$$

and

$$\frac{\partial \alpha}{\partial \mathbf{x}} = \left(\frac{\partial \mathbf{e}_2}{\partial \mathbf{x}}\right) \cdot \mathbf{e}_1 + \left(\frac{v_{\parallel}}{v_{\perp}^2}\right) \frac{\partial}{\partial \mathbf{x}} \mathbf{e}_{\parallel} \cdot (\mathbf{v}_{\perp} \times \mathbf{e}_{\parallel}) \quad , \tag{IV.6.5}$$

with $\mathbf{e}_1, \mathbf{e}_2$ and $\mathbf{e}_{\parallel}$ being local orthogonal unit vectors. Meanwhile, the Vlasov propagator becomes, assuming $\mathbf{E}_o = 0$,

$$\mathcal{L}_o + \mathcal{L}_g = \frac{\partial}{\partial t} + v_{\parallel} \frac{\partial}{\partial X_{\parallel}} + \mathbf{v} \cdot (\boldsymbol{\lambda}_{B1} + \boldsymbol{\lambda}_{B2}) - \Omega \frac{\partial}{\partial \alpha} \quad . \tag{IV.6.6}$$

Consider the equilibrium distribution function, f_{og}, which obeys

$$\Big[\overset{\eta}{v_{\parallel} \frac{\partial}{\partial X_{\parallel}}} + \mathbf{v} \cdot (\overset{\eta^2}{\boldsymbol{\lambda}_{B1}} + \overset{\eta}{\boldsymbol{\lambda}_{B2}}) - \overset{1}{\Omega \frac{\partial}{\partial \alpha}}\Big] f_{og}(\mathbf{V}, \mathbf{X}) = 0 \quad . \tag{IV.6.7}$$

For O(1), we have $\partial f_{og}/\partial \alpha = 0$; i.e.,

$$f_{og} = f_{og}(\varepsilon, \mu, \sigma, \mathbf{X}) \quad . \tag{IV.6.8}$$

Now, for the sake of simplicity, we shall further assume f_{og} is <u>isotropic</u>; i.e., $f_{og} = f_{og}(\varepsilon, \mathbf{X})$. Then, for $O(\eta)$, we obtain

$$v_\parallel \frac{\partial}{\partial X_\parallel} f_{og} = \Omega \frac{\partial}{\partial \alpha} f_{og}^{(1)} \quad . \tag{IV.6.9}$$

Gyroaveraging over Eq. (IV.6.9) then yields $\partial f_{og}/\partial X_\parallel = 0$; i.e.,

$$f_{og} = f_{og}(\varepsilon, \underset{\sim}{X}_\perp) \quad . \tag{IV.6.10}$$

As to the perturbations, we shall only consider the electrostatic limit. Thus letting

$$\delta f_g = \frac{q}{m} \delta\phi_g \frac{\partial f_{og}}{\partial \varepsilon} + \delta G_g \quad , \tag{III.6.11}$$

we find, recalling the orderings given by Eq. (III.7.1),

$$\begin{aligned} \pounds_g \delta G_g &= \Big[\overset{\eta}{\frac{\partial}{\partial t}} + v_\parallel \overset{\eta}{\frac{\partial}{\partial X_\parallel}} + \underset{\sim}{v} \cdot \big(\overset{\eta}{\underset{\sim}{\lambda}_{B1}} + \overset{\eta}{\underset{\sim}{\lambda}_{B2}}\big) - \Omega \overset{1}{\frac{\partial}{\partial \alpha}}\Big] \delta G_g \\ &= - \frac{q}{m} \Big(\frac{\partial \delta\phi_g}{\partial t} \overset{\eta}{\frac{\partial}{\partial \varepsilon}} - \frac{\partial}{\partial \underset{\sim}{X}} \delta\phi_g \times \overset{\eta}{\frac{\underset{\sim}{e}_\parallel}{\Omega}} \cdot \frac{\partial}{\partial \underset{\sim}{X}}\Big) f_{og} \quad . \end{aligned} \tag{IV.6.12}$$

Expanding $\delta G_g = \delta G_{go} + \delta G_{g1} + \ldots$, we find, for $O(1)$,

$$\frac{\partial \delta G_{go}}{\partial \alpha} = 0 \quad . \tag{IV.6.13}$$

Gyroavering the $O(\eta)$ equation, we obtain

$$\langle \pounds_g \rangle_o \, \delta G_{go} = - \frac{q}{m} \Big(\frac{\partial}{\partial t} \langle \delta\phi_g \rangle_0 \frac{\partial}{\partial \varepsilon} - \frac{\partial}{\partial \underset{\sim}{X}} \langle \delta\phi_g \rangle_o \times \frac{\underset{\sim}{e}_\parallel}{\Omega} \cdot \frac{\partial}{\partial \underset{\sim}{X}}\Big) f_{og} \quad , \tag{IV.6.14}$$

where, correct to $O(\eta)$,

$$\langle \mathcal{L}_g \rangle_o = \frac{\partial}{\partial t} + v_\parallel \frac{\partial}{\partial X_\parallel} + \underset{\sim}{v}_d \cdot \frac{\partial}{\partial \underset{\sim}{X}} \quad , \tag{IV.6.15}$$

$$\underset{\sim}{v}_d \cdot \frac{\partial}{\partial \underset{\sim}{X}} = \langle \underset{\sim}{v} \cdot \underset{\sim}{\lambda}_{B1} \rangle_0 \quad , \tag{IV.6.16}$$

and

$$\underset{\sim}{v}_d = \frac{\underset{\sim}{e}_\parallel \times [(v_\perp^2/2)\underset{\sim}{\nabla}_X \ln B_o + v_\parallel^2 \underset{\sim}{e}_\parallel \cdot \underset{\sim}{\nabla}_X \underset{\sim}{e}_\parallel]}{\Omega} \quad . \tag{IV.6.17}$$

Here, $\underset{\sim}{v}_d$ corresponds to the curvature and $\underset{\sim}{\nabla}B$ drifts of the guiding center. Also, in deriving Eq. (IV.6.15), we note that $\langle \partial\mu/\partial\underset{\sim}{X} \rangle_o = O(\eta^2)$. Equation (IV.6.14) is the desired linear gyrokinetic equation valid for nonuniform magnetized plasmas.

Let us go further with the following eikonal approximation in the perpendicular to $\underset{\sim}{B}_o$ direction, i.e.,

$$\delta f_g = \delta \hat{f}_g(\underset{\sim}{X}) \exp[i(\int^{\underset{\sim}{X}_\perp} \underset{\sim}{k}_\perp \cdot d\underset{\sim}{X}_\perp - \omega t)] \quad . \tag{IV.6.18}$$

Equation (IV.6.14) then becomes

$$\left(v_\parallel \frac{\partial}{\partial X_\parallel} - i\omega + i\underset{\sim}{k}_\perp \cdot \underset{\sim}{v}_d\right)\delta\hat{G}_{go} = -i(\hat{\omega} + \omega_*) f_{og} \, J_o \delta\hat{\phi} \, \frac{q}{T} \quad , \tag{IV.6.19}$$

where we have denoted $\hat{\omega} = -\,\omega\varepsilon_o \partial/\partial\varepsilon$, $T = m\varepsilon_o$,

$$\frac{\partial f_{og}}{\partial \underset{\sim}{X}_\perp} = -\,\underset{\sim}{\kappa}\, f_{og} \quad , \tag{IV.6.20}$$

$$\omega_* = \frac{Tc}{qB_o} \underset{\sim}{k}_\perp \times \underset{\sim}{e}_\parallel \cdot \underset{\sim}{\kappa} \quad , \tag{IV.6.21}$$

and have assumed no temperature gradient, $\partial T/\partial \underset{\sim}{X}_\perp = 0$.

§IV.7 Drift instabilities in eikonal approximation

Here, we adopt the slab plasma model. Thus, $\underset{\sim}{B}_0 = B_0 \underset{\sim}{e}_z$, $f_{og} = N_0(X) f_M$, and, with $\beta \ll 1$ assumed, $\underset{\sim}{B}_0$ is taken to be uniform. Assuming perturbations of the form [c.f. Eq. (IV.6.18)]

$$\delta\hat{f}_g(\underset{\sim}{X}) = \delta\bar{f}_g \exp(i\, k_\parallel\, Z) \quad , \tag{IV.7.1}$$

Eq. (IV.6.19) then gives, with $\underset{\sim}{k}_\perp = k_x(x)\underset{\sim}{e}_x + k_y\, \underset{\sim}{e}_y$,

$$(\omega - k_\parallel v_\parallel)\overline{\delta G_{go}} = \frac{q}{T}\,(\omega + \omega_*) N_0(X) J_0 \overline{\delta\phi} \quad . \tag{IV.7.2}$$

For electrons, we have $|k_\perp \rho_e| \ll 1$ and, hence, $\delta\bar{f}_e = \delta\bar{f}_{ge}$. That is, we find

$$\frac{\delta\bar{n}_e}{N_0} = \frac{e}{T_e}\,\overline{\delta\phi} + \frac{1}{N_0}\int d^3\underset{\sim}{v}\ \delta\bar{G}_{eo} = \frac{e}{T_e}\,\delta\bar{\phi}\Big[1 + \Big(1 - \frac{\omega_{*e}}{\omega}\Big)\xi_e Z_e\Big] \quad . \tag{IV.7.3}$$

Here, $\xi_e = \omega/|k_\parallel v_{te}|$ and $Z_e = Z(\xi_e)$. As to ions, we find

$$\frac{\delta\bar{n}_i}{N_0} = -\frac{e}{T_i}\,\delta\bar{\phi} + \frac{1}{N_0}\int d^3\underset{\sim}{v}\ J_{oi}\ \delta\bar{G}_{goi} = -\frac{e}{T_i}\,\delta\bar{\phi}\Big[1 + \Big(1 + \frac{\omega_{*i}}{\omega}\Big)\Gamma_{oi}\xi_i Z_i\Big] \quad . \tag{IV.7.4}$$

Substituting $\delta\bar{n}_e$ and $\delta\bar{n}_i$ into the quasi-neutrality condition, we obtain the following linear dispersion relation

$$1 + \tau(1-\Gamma_{oi}) + \Big(1-\frac{\omega_{*e}}{\omega}\Big)\xi_e Z_e + \frac{\omega_{*e}}{\omega}\,\Gamma_{oi}\xi_i Z_i + \tau\Gamma_{oi}(1+\xi_i Z_i) = 0 \quad . \tag{IV.7.5}$$

Here, $\tau \equiv T_e/T_i$ and we have noted that $\tau\omega_{*i} = \omega_{*e}$.

To further analyze Eq. (IV.7.5), let us assume $|\xi_i| >> 1$, $\tau >> 1$ and $k_\perp^2\rho_s^2 \lesssim 1$. Equation (IV.7.5) then becomes approximately

$$\omega^2(1 + k_\perp^2\rho_s^2) - \omega\omega_{*e}(1 - b_i) + \omega(\omega - \omega_{*e})\xi_e Z_e - k_\parallel^2 c_s^2 = 0 \quad . \qquad \text{(IV.7.6)}$$

Equation (IV.7.6) shows that in nonuniform plasmas, the electrostatic ion-acoustic mode is modified by the diamagnetic drifts. To see that such a modification can lead to instability, we further assume that $|\omega| \sim |\omega_{*e}| > |k_\parallel c_s|$, $|\xi_e| << 1$, $\omega = \omega_r + i\gamma$ and $|\gamma/\omega_r| << 1$. We than derive, from Eq. (IV.7.6),

$$\omega_r \simeq \frac{\omega_{*e}(1-b_i)}{1+k_\perp^2\rho_s^2} \quad , \qquad \text{(IV.7.7)}$$

and

$$\frac{\gamma}{\omega_{*e}} \simeq \sqrt{\pi}\, \xi_e \left(\frac{k_\perp\rho_s}{1+k_\perp^2\rho_s^2}\right)^2 \quad . \qquad \text{(IV.7.8)}$$

Note the instability predicted by Eqs. (IV.7.7) and (IV.7.8) requires only the presence of density gradient, finite ion Larmor radius and electron Landau damping; all of which are universal ingredients in magnetically confined plasmas. This instability, therefore, is often called the universal drift instability.

§IV.8 Convective damping of drift waves in a sheared magnetic field: an eigenmode analysis

[Ref. Pearlstein and Berk, Phys. Rev. Lett. (1969)]

It often occurs that stability analyses via the eikonal assumption break down due to the presence of turning points and, hence, the associated wave reflection. In such cases, the stability analysis take the form of eigenvalue problems with the frequencies, ω, being the eigenvalues. We shall illustrate this kind of eigenmode stability analysis using the well-known example of the convective damping of drift waves in a magnetic field with finite shear or, simply, the shear damping of drift waves.

Again, we adopt the slab-plasma model considered in Sec. §IV.7. This time, however, the magnetic field is sheared, i.e.,

$$\underset{\sim}{B} = B_o\left(\underset{\sim}{e}_z + \frac{x}{L_s}\underset{\sim}{e}_y\right) \quad , \tag{IV.8.1}$$

where L_s is the shear scale length. Considering perturbations of the form

$$\delta\phi(\underset{\sim}{x},t) = \delta\bar{\phi}(x)\,\exp[i(k_y y - \omega t)] \quad , \tag{IV.8.2}$$

we then have $k_\parallel = \underset{\sim}{k}\cdot\underset{\sim}{B}/B \simeq k_y x/L_s \equiv k_\parallel' x$ with $O(x^2/L_s^2)$ corrections ignored. Since $k_\parallel = 0$ at $x = 0$, $x = 0$ is called the mode-rational surface.

To further simplify the analysis, we shall assume ions are sufficiently cold such that $T_e \gg T_i$ and $|k_\perp \rho_i|^2$, $|k_\parallel v_i/\omega|^2 \ll 1$. Furthermore, we shall suppress electron-wave resonance by letting $|\omega/k_\parallel v_e| \to 0$ so that we may concentrate on the shear-induced convective damping. Under these assumptions, the corresponding eigenmode equation can be readily obtained from Eq. (IV.7.6), with the $\xi_e Z_e$ term neglected, i.e.,

$$\left(\rho_s^2 \frac{d^2}{dx^2} + E - V(x)\right)\delta\bar{\phi}(x) = 0 \quad , \tag{IV.8.3}$$

where

$$E = \frac{\omega_{*e}}{\omega} - 1 - k_y^2\rho_s^2 \quad , \tag{IV.8.4}$$

and

$$V(x) = -(\frac{k_\parallel' x c_s}{\omega})^2 \quad . \tag{IV.8.5}$$

Note that we have casted Eq. (IV.8.3) in the form of a Schrödinger equation. Thus, with $\omega = \omega_r + i\omega_i$ and assuming $|\omega_r| >> |\omega_i|$, this eigenmode equation corresponds to a harmonic oscillator (Weber equation) with a parabolic potential hill, or, sometimes called, anti-well [c.f. Fig. (IV.8.1)].

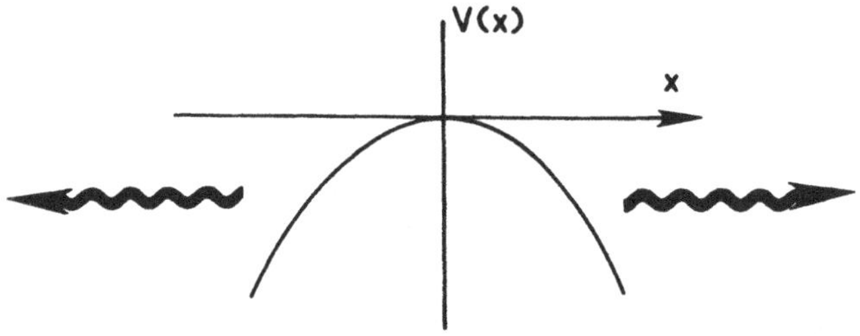

Fig. (IV.8.1) Sketch of the "anti" potential well V(x) of Eq. (IV.8.5). Here, the wave energy leaks away from the system.

Now, as an eigenvalue problem, we need to specify the boundary conditions. As $|x| \to \infty$, V(x) term dominates and we have wave propagation; i.e., waves can propagate either into or away from the system. The proper way to determine the boundary conditions is then to regard the eigenmode analysis as the time-asymptotic limit of an initial-value problem. In this respect, we must apply the causality requirement; that is,

$$\lim_{|x|\to\infty} |\delta\bar{\phi}(x)| \to 0 \quad \text{for} \quad \text{Im}\,\omega > 0 \quad . \tag{IV.8.6}$$

Noting that, as $|x| \to \infty$, $\delta\bar{\phi}(x)$ has the following WKB solutions

$$\lim_{|x|\to\infty} \delta\bar{\phi}(x) \to i \exp\left(\pm i \frac{k_\parallel' c_s x^2}{2\omega\rho_s}\right) \quad . \tag{IV.8.7}$$

Imposing the causality condition, Eq. (IV.8.6), then rejects the solution in Eq. (IV.8.7) with the plus sign. The proper boundary condition is, thus, given by

$$\lim_{|x|\to\infty} \delta\phi(x) \to \exp\left(- i \frac{k_\parallel' c_s x^2}{2\omega\rho_s}\right) \quad . \tag{IV.8.8}$$

We now examine the physical implication of this boundary condition to the wave propagation. In terms of WKB solutions of the form, $\exp(-i\omega t + i \int_x^x k_x dx)$, Eq. (IV.8.8) corresponds to

$$k_x = - \frac{k_\parallel' c_s x}{\omega \rho_s} \quad , \tag{IV.8.9}$$

or

$$\omega = - \frac{k_\parallel' c_s x}{\rho_s k_x} \quad . \tag{IV.8.10}$$

We then obtain

$$v_{gx} = \frac{\partial\omega}{\partial k_x} = \frac{k_\parallel' c_s x}{k_x^2 \rho_s} \quad ; \tag{IV.8.11}$$

i.e., $v_{gx} > 0$ for $x > 0$ and $v_{gx} < 0$ for $x < 0$ [c.f. Fig. IV.8.1)]. That is, the wave energy is <u>leaking</u> away from the system. Looking from the initial-value point of view, and, in the absence of any instability source feeding the waves, any initial perturbations can, thus, be expected to decay in time due to the convective wave energy leakage. This constitutes the physical picture of the shear damping.

In the followings, we perform the eigenmode analysis to demonstrate this shear-damping mechanism. From the asymptotic solution, Eq. (IV.8.8) we know the asymptotic anti-Stokes lines are given by

$$2\theta_x = \begin{cases} \theta_\omega,\ \theta_\omega - \pi \quad \text{for} \quad \text{Re } x > 0 \quad ; \\ \theta_\omega + \pi,\ \theta_\omega + 2\pi \quad \text{for} \quad \text{Re } x < 0 \quad , \end{cases} \qquad \text{(IV.8.12)}$$

where $\theta_x = \arg(x)$ and $\theta_\omega = \arg(\omega)$. In Fig. (IV.8.2), we have sketched these lines assuming Im $\omega > 0$ (i.e., $\theta_\omega > 0$).

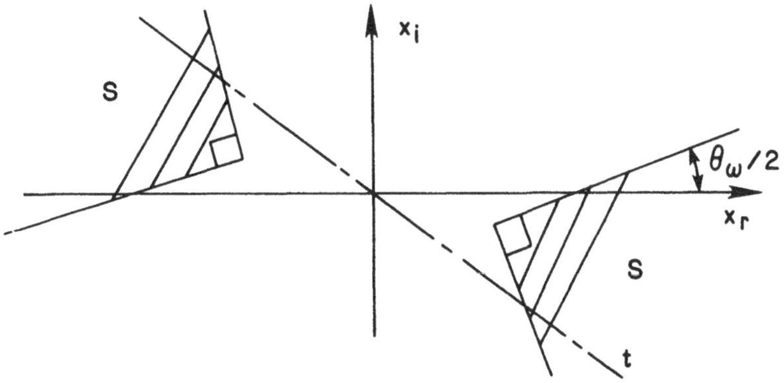

Fig. (IV.8.2) Asymptotic Stokes structures of Eq. (IV.8.8) and the transformed coordinate t given by Eqs. (IV.8.13) and (IV.8.14).

Furthermore, from the causality condition, Eq. (IV.8.6), the solution must decay along the real-x(x_r) axis and, hence, be subdominant in the hatched (S) regions. We can then make a transformation to the t-axis defined by

$$x = \alpha t \quad , \qquad \text{(IV.8.13)}$$

and

$$\alpha = \left(\frac{\omega\rho_s}{k_\parallel' c_s}\right)^{1/2} \exp\left(-\frac{i\pi}{4}\right) \quad , \tag{IV.8.14}$$

such that the t-axis lies inside the <u>subdominant</u> regions. The eigenmode equation , Eq. (IV.8.3), then becomes

$$\left(\frac{d^2}{dt^2} + \frac{E\alpha^2}{\rho_s^2} - t^2\right)\delta\bar{\phi}(t) = 0 \quad . \tag{IV.8.15}$$

The eigenvalue condition is then

$$\frac{E\alpha^2}{\rho_s^2} = (2n + 1) \text{ for } n = 0, 1, \ldots \quad ; \tag{IV.8.16}$$

that is, noting Eqs. (IV.8.4) and (IV.8.14), the dispersion relation is

$$\frac{\omega}{\omega_{*e}} = \frac{1}{1+k_y^2\rho_s^2} - i\,\frac{(2n+1)}{1+k_y^2\rho_s^2}\,\frac{L_n}{L_s} \quad . \tag{IV.8.17}$$

Hence, the eigenmodes are indeed <u>damped</u> by the finite magnetic shear.

Finally, we remark that our analysis is valid even if the obtained ω has $\text{Im}\,\omega < 0$. This is because, from the argument of analytic continuation, the hatched regions bounded by the asymptotic anti-Stokes lines [c.f. Fig. (IV.8.2)] remain to be subdominant as ω is continuated from the upper-ω to the lower-ω half plane. In other words, the boundary condition, Eq. (IV.8.8), is solely determined by the causality condition and remains unchanged throughout the analytic continuation in the ω plane.

CHAPTER V

Topics in Nonlinear Plasma Theory

§V.1 Quasi-linear theory

In previous chapters, we have shown, via linear theories, that there exist various mechanisms to either excite or damp waves in plasmas. In other words, plasmas as dielectric media may either emit or absorb wave packets (quasi particles). Considering the plasma-quasi particle system as a whole, we would expect some types of energy and momentum conservation and, therefore, that the plasma will react to the emission and/or absorption process. This subject is, of course, beyond the linear theory and is called, correspondingly, the quasi-linear theory.

In the followings, we list some characteristics of the quasi-linear theory:

(i) It deals with effects of either spontaneously or externally excited perturbed (microscopic) fluctuations on "equilibrium" (macroscopic) quantities.

(ii) The perturbed fluctuations are taken to be of suffiently small amplitudes such that perturbative expansions are justified.

(iii) It ignores nonlinear mode-coupling processes (e.g. parametric decays).

(iv) In reaction to the perturbations, the "equilibrium" quantities evolve slowly (adiabatically) in time.

In these lectures, we shall only consider the Vlasov system in infinite, uniform plasmas. Thus, in this case, the corresponding physical quantity is the velocity distribution function. For simplicity, we further limit to the one-dimensional electrostatic case. [Ref. A. N. Kaufman, J. Plasma Phys. **8**, 1 (1972).] Generalization to higher-dimensional, magnetized and electromagmetic cases are conceptually straightforward. In fact, we could also generalize the underlying principles to nonuniform plasmas. We, therefore, can expect quasi-linear evolution of "equilibrium" pressure and/or current profiles and, in this sense, quasi-linear theories have important applications in understanding anomalous transport processes and/or nonlinear saturation of kink/tearing modes.

Now, in the one-dimensional electrostatic case, the Vlasov equation is given by

$$\left(\frac{\partial}{\partial t} + v\,\frac{\partial}{\partial x} + \frac{q}{m}\,\delta E\,\frac{\partial}{\partial v}\right)f = 0 \quad . \tag{V.1.1}$$

For definitiveness, let us consider the plasma to be periodic with a large periodicity length, L. (Later, we shall take $L \to \infty$ to reach the continuous limit). We, thus, may separate f into its equilibrium and perturbed parts; i.e.,

$$f(x,\ v,\ t) = \langle f\rangle_x + \delta f \quad , \tag{V.1.2}$$

where

$$\langle f\rangle_x \equiv \left(\frac{1}{L}\right)\int_0^L dx\ f \quad , \tag{V.1.3}$$

and

$$\left|\frac{\delta f}{\langle f\rangle_x}\right| \sim O(\eta) \ll 1 \qquad (V.1.4)$$

is assumed. Here, η is the smallness parameter ordering the perturbations. Spatial-averaging Eq. (V.1.1), we obtain

$$\frac{\partial}{\partial x}\langle f\rangle_x = -\frac{q}{m}\frac{\partial}{\partial v}\langle \delta E \delta f\rangle_x \quad . \qquad (V.1.5)$$

Noting that the right hand side of Eq. (V.1.5) is of $O(\eta^2)$, $\langle f\rangle_x$, thus, indeed evolves slowly in time. Subtracting Eq. (V.1.5) from Eq. (V.1.1), we have

$$\overset{\eta}{\left(\frac{\partial}{\partial t} + v\frac{\partial}{\partial x}\right)\delta f} + \overset{\eta}{\frac{q}{m}\delta E\frac{\partial}{\partial v}\langle f\rangle_x} + \overset{\eta^2}{\frac{q}{m}\frac{\partial}{\partial v}\left(\delta E \delta f - \langle \delta E \delta f\rangle_x\right)} = 0 \quad . \qquad (V.1.6)$$

In addition, we have the Poisson's equation to complete the system of equations; i.e.,

$$\frac{\partial \delta E}{\partial x} = 4\pi q n_0 \int dv\, \delta f \quad , \qquad (V.1.7)$$

where $n_0 = N/L$, N is the total number of particles, and a single-species plasma is assumed here for the sake of simplicity.

Noting that $\partial\langle f\rangle_x/\partial t \sim O(\eta^2)$, we can express $\langle f\rangle_x$ as

$$\langle f\rangle_x = f_0(v, \eta^2 t) + f_2(v, t) \quad . \qquad (V.1.8)$$

Furthermore, as mentioned above, we shall ignore the nonlinear mode-coupling effects; i.e., the $O(\eta^2)$ terms in Eq. (V.1.6). We, thus, have

$$\left(\frac{\partial}{\partial t} + v\frac{\partial}{\partial x}\right)\delta f = -\frac{q}{m}\delta E\frac{\partial}{\partial v} f_0 \quad , \qquad (V.1.9)$$

or

$$\delta f(x,v,t) = \delta f_0(x-vt,v,0) - \frac{q}{m}\int_0^t d\tau\ \delta E(x-v\tau,\ t-\tau)f_0'(v,t-\tau) \quad . \qquad (V.1.10)$$

Expressing the perturbed electric field δE as

$$\delta E(x,t) = \sum_k \delta\hat{E}_k(\eta t)\exp(ikx - i\omega_k t) \qquad (V.1.11)$$

with $k = 2n\pi/L$ and n = nonzero integers, the Poisson's equation, Eq. (V.1.7), becomes

$$ik\delta\hat{E}_k = 4\pi n_0 q\Big(\int dv\ \delta f_{0k}\ \exp[i(\omega_k - kv)t] - \frac{q}{m}\int_0^t d\tau\ \delta\hat{E}_k(t-\tau)\exp(i\omega_k\tau)$$

$$\times \int dv\ \exp(-ikv\tau)f_0'(v;\ t-\tau)\Big) \quad . \qquad (V.1.12)$$

From our discussions in Chapter II on the phase-mixing of ballistic modes, it is clear that, given sufficiently analytic and broad-band δf_{0k} and f_0, the velocity integrations in Eq. (V.1.12) will yield contributions which decay in time scales of $O(1)$; i.e., much shorter compared to the $O(1/\eta)$ time scale of $\delta\hat{E}_k$. Equation (V.1.12) can, therefore, be approximated as

$$ik\ \delta\hat{E}_k \simeq -\omega_p^2 \int dvf_0'(v;t) \int_0^\infty d\tau[\delta\hat{E}_k(t) - \tau\delta\dot{\hat{E}}_k]\ \exp[i(\omega_k - kv)\tau] \quad . \qquad (V.1.13)$$

Using the causality condition (i.e., Im $\omega_k \rightarrow 0^+$), we have

$$\int_0^\infty d\tau\ \exp[i(\omega_k - kv)\tau] = iP\left(\frac{1}{\omega_k - kv}\right) + \pi\delta(kv - \omega_k) \equiv \pi\delta_+ \quad . \qquad (V.1.14)$$

Equation (V.1.12) then can be expressed as

$$\varepsilon_{kr} + i\varepsilon_{ki} = -i\gamma_k \frac{\partial\varepsilon_{kr}}{\partial\omega_k} \quad , \tag{V.1.15}$$

where

$$\varepsilon_{kr}(t) = 1 + \frac{\omega_p^2}{k} \int dv\ f_o'(v;t) P\left(\frac{1}{\omega_k - kv}\right) \quad , \tag{V.1.16}$$

$$\varepsilon_{ki}(t) = -\frac{\omega_p^2}{k} \pi \int dv\ f_o'(v;t) \delta(kv - \omega_k) \quad , \tag{V.1.17}$$

and

$$\gamma_k \equiv \frac{\delta\dot{\hat{E}}_k}{\delta\hat{E}_k} \quad . \tag{V.1.18}$$

Equation (V.1.15) is, of course, just the usual linear dispersion relation. Thus, within the framework of the quasi-linear theory, the perturbations still satisfy the standard linear dispersion relation; which, however, evolves slowly in time since f_o does.

With the linear perturbed responses determined, we now proceed to calculate the time evolution of $\langle f \rangle_x$ given by Eq. (V.1.5). To this end, we need to calculate $\langle \delta E \delta f \rangle_x$ which, using Eq. (V.1.10) and the periodicity condition, can be shown to be

$$\langle \delta E \delta f \rangle_x = \sum_k \delta\hat{E}_{-k}\ \delta f_{ok} \exp[i(\omega_k - kv)t] - \frac{q}{m} \int_o^t d\tau \sum_k \delta\hat{E}_k(t) \delta\hat{E}_k(t-\tau)$$

$$\times \exp[i(\omega_k - kv)\tau] f_o'(v;t-\tau) \quad . \tag{V.1.19}$$

Now, if the perturbations, $\delta\hat{E}_k$ and δf_{ok}, are sufficiently broad in the k spectrum; e.g. let us assume they peak at $k = k_o$ and have a width Δk, we can then argue that the following expression

$$\sum_k |\delta\hat{E}_k|^2 \exp[i(\omega_k - kv)\tau]$$

decay on a characteristic time scale

$$\tau_o \equiv \left[\Delta k\left(v - \frac{d\omega}{dk}\right)\Big|_{k = k_o}\right]^{-1} \quad . \tag{V.1.20}$$

τ_o, thus, is formally of O(1) and, in the long time scale, Eq. (V.1.19) can be further approximated as

$$\langle\delta E\delta f\rangle_x \simeq -\frac{q}{m} f_o'(v;t) \int_0^\infty d\tau \sum_k \left(|\delta\hat{E}_k|^2 - \frac{\tau}{2}\frac{d}{dt}|\delta\hat{E}_k|^2\right)\exp[i(\omega_k - kv)\tau]$$

$$= f_o'\left(\pi \sum_k |\delta\hat{E}_k|^2 \delta_+ + i\,\frac{\pi}{2}\frac{d}{dt}\sum_k |\delta\hat{E}_k|^2 \frac{\partial}{\partial\omega_k}\delta_+\right) \quad ; \tag{V.1.21}$$

where we have utilized Eq. (V.1.14).

To be consistent with Kaufman's notations, we shall first symmetrize $|\delta\hat{E}_k|^2$ with respect to $k = 0$. Furthermore, we take $L \to \infty$ and approach the continuous limit. Correspondingly, we have

$$\sum_k |\delta\hat{E}_k|^2 \to \int_0^\infty \frac{dk}{2\pi}\,\varepsilon(k,t) \quad , \tag{V.1.22}$$

and Eq. (V.1.5) along with (V.1.8) becomes

$$\frac{\partial f_o}{\partial t} + \frac{\partial f_2}{\partial t} = \frac{q^2}{m^2}\frac{\partial}{\partial v}(I\ f_o') \quad , \qquad (V.1.23)$$

where, noting Eq. (V.1.21), I is given by

$$I = I_o + \frac{dI_1}{dt} \quad , \qquad (V.1.24)$$

$$I_o = \int_o^\infty \frac{dk}{2\pi}\ \varepsilon\ 2\pi\delta(\omega_k - kv) \quad , \qquad (V.1.25)$$

and

$$I_1 = -\int_o^\infty \frac{dk}{2\pi}\ \varepsilon\ \frac{\partial}{\partial\omega_k} P\left(\frac{1}{\omega_k - kv}\right) \quad . \qquad (V.1.26)$$

Alternatively, we may express Eq. (V.1.23) as

$$\frac{\partial f_o}{\partial t} = \frac{q^2}{m^2}\frac{\partial}{\partial v}(I_o f_o') = \frac{\partial}{\partial v}\left(\tilde{D}_r \frac{\partial f_o}{\partial v}\right) \quad , \qquad (V.1.27)$$

$$f_2 = \frac{q^2}{m^2}\frac{\partial}{\partial v}\left(I_1 \frac{\partial f_o}{\partial v}\right) \quad , \qquad (V.1.28)$$

with $\tilde{D}_r = (q/m)^2 I_o$. Since only resonant particles with $v = \omega_k/k$ contribute to I_o, Eq. (V.1.27) describes the time evolution of the resonant particles; which, from the form of the equation, are diffusing in the velocity space with $\tilde{D}_r$ being the diffusion coefficient. Meanwhile, f_2 corresponds to the quasi-linear deformation of non-resonant particles since they do jiggle under the influence of waves.

We now establish some conservation properties. First, it is trivial to show that the number of particles is conserved; i.e.,

$$\frac{d}{dt} \int dv \, \langle f \rangle_x = \int dv \, \frac{\partial}{\partial t} \left(f_0 + f_2 \right) = 0 \quad . \tag{V.1.29}$$

Next, we prove the conservation of momentum, i.e.,

$$\frac{d}{dt} \sum_j n_{oj} m_j \int dv \, v \langle f_j \rangle_x = 0 \quad . \tag{V.1.30}$$

Here, we do the simple generalization to multi-species. Noting that

$$\frac{d}{dt} \sum_j n_{oj} m_j \int dv \, v f_{oj} = - \sum_j n_{oj} m_j \int dv \, \tilde{D}_{rj} \frac{\partial f_{oj}}{\partial v}$$

$$= - \sum_j \int_0^\infty \frac{dk}{2\pi} \int dv \, \omega_{pj}^2 \frac{\varepsilon}{4\pi} 2\gamma_k f'_{oj} \frac{\partial}{\partial \omega_k} P\left(\frac{1}{\omega_k - kv}\right) \; ; \tag{V.1.31}$$

where we have used the linear dispersion relation, Eq. (V.1.15), and that

$$\frac{d}{dt} \sum_j n_{oj} m_j \int dv \, v f_{2j} = - \sum_j n_{oj} \left(\frac{q}{m}\right)_j^2 \int dv \, f'_{oj} \frac{d}{dt} I_{1j}$$

$$= \sum_j \omega_{pj}^2 \int dv \int_0^\infty \frac{dk}{2\pi} f'_{oj} 2\gamma_k \frac{\varepsilon}{4\pi} \frac{\partial}{\partial \omega_k} P\left(\frac{1}{\omega_k - kv}\right) , \tag{V.1.32}$$

Eq. (V.1.30) is thereby proved. Following the same strategy, we can readily prove that the total energy is also conserved, i.e.,

$$\frac{d}{dt} \left(\sum_j \frac{1}{2} n_{oj} m_j \int dv \, v^2 \langle f_j \rangle_x + \int_0^\infty \frac{dk}{2\pi} \frac{\varepsilon}{4\pi} \right) = 0 \quad . \tag{V.1.33}$$

Finally, let us examine the time-asymptotic behavior of the f_0. Multiplying Eq. (V.1.27) by f_0 and integrating over the velocity, we obtain

$$\frac{1}{2}\frac{d}{dt}\int dv\ f_0^2 = -\int dv\ \tilde{D}_r\left(\frac{\partial f_0}{\partial v}\right)^2 \quad . \tag{V.1.34}$$

Now, $\tilde{D}_r \geq 0$ and, hence, we have

$$\frac{1}{2}\frac{d}{dt}\int dv\ f_0^2 = -\int dv\ \tilde{D}_r\left(\frac{\partial f_0}{\partial v}\right)^2 \leq 0 \quad . \tag{V.1.35}$$

However, with f_0 being a physical quantity, we also require

$$\int dv\ f_0^2 > 0 \quad \text{for all } t \quad . \tag{V.1.36}$$

Equations (V.1.35) and (V.1.36) then uniquely impose the following time-asymptotic condition; i.e.,

$$\lim_{t\to\infty}\int dv\ \tilde{D}_r\left(\frac{\partial f_0}{\partial v}\right)^2 \to 0 \quad . \tag{V.1.37}$$

Now, in the case of damped perturbations, Eq. (V.1.37), of course, is easily satisfied since $\varepsilon \to 0$ as $t \to \infty$. In the cases of <u>unstable</u> or <u>undamped</u> (e.g. externally driven) perturbations, however, we have $\tilde{D}_r > 0$ for the resonant particles ($v = v_r \equiv \omega_k/k$). Equation (V.1.37), then indicates that

$$\lim_{t\to\infty}\left.\frac{\partial f_0}{\partial v}\right|_{v-v_r} \to 0 \quad ; \tag{V.1.38}$$

i.e., the velocity distribution becomes flattened at the resonant velocities. This is the well-known quasi-linear flattening of the resonant-particle velocity distribution.

§V.2 Ponderomotive force

In this section, we shall consider the long-time single-particle motion in an electromagnetic field. Here, the meaning of "long-time" is relative to the periods of the wave oscillations. Thus, as may be expected, the analysis employs two time scales and involves the phase averaging of the fast time scales.

To simplify the analysis, let us consider the waves to be electrostatic and the frequencies are peaked near ω_0. The corresponding single-particle equation of motion is then

$$\ddot{x} = (\frac{q}{m})[\delta\hat{E}(\eta t,x)\exp(-i\omega_0 t) + c.c.] \equiv (\frac{q}{m})(\delta E + \delta E^*) \quad . \tag{V.2.1}$$

Here, we have further limited to the one-dimensional case. Assuming the wave amplitude is sufficiently small, we can then carry out a perturbative expansion; i.e., we let

$$x = x_0 + x_1 + \ldots \quad . \tag{V.2.2}$$

In the zeroth order, we find $\ddot{x}_0 = 0$ or

$$x_0 = c_0 + v_0 t \quad . \tag{V.2.3}$$

x_0, of course, is just the particle's unperturbed orbit. In the first order, we have

$$\ddot{x}_1 = (\frac{q}{m})[\delta\hat{E}(\eta t,x_0)\exp(-i\omega_0 t) + c.c.]$$

$$= \left(\frac{q}{m}\right)\left[\exp(-i\omega_k t)\sum_k \delta\hat{E}_k(\eta t)\exp(ikx_o) + c.c.\right] \quad . \qquad (V.2.4)$$

Here, we shall make the following crucial assumption that the waves have <u>high</u> phase velocities; i.e.,

$$\left|\frac{\omega_o}{k}\right| \gg v \quad ; \qquad (V.2.5)$$

which, in this case, corresponds to the limit that the number of resonant particles is negligible compared to that of non-resonant particles or, equivalently, that the waves are either weakly damped or weakly growing. Under this assumption, we find

$$x_1 = -\left(\frac{q}{m}\,\omega_o^2\right)(\delta E_o + \delta E_o^*) \equiv X_1 + X_1^* \quad , \qquad (V.2.6)$$

with

$$\delta E_o \equiv \delta\hat{E}(\eta t, c_o)\exp(-i\omega_o t) \quad . \qquad (V.2.7)$$

x_1, thus, corresponds to the jiggling motion (oscillation) about x_o. Proceed further to the second order, we have

$$\ddot{x}_2 = \left(\frac{q}{m}\right) x_1\left(\frac{\partial \delta E_o}{\partial x_o} + \frac{\partial \delta E_o^*}{\partial x_o}\right) \quad . \qquad (V.2.8)$$

Since we are interested in the long-time dynamics, we shall average out the fast oscillations with period $T_o \equiv 2\pi/\omega_o$. Thus, defining

$$\langle \ldots \rangle_o = \frac{1}{T_o}\int_o^{T_o} dt(\ldots) \quad , \qquad (V.2.9)$$

we find

$$\langle \ddot{x}_2 \rangle_0 = \frac{q}{m} \langle X_1^* \frac{\partial \delta E_0}{\partial x_0} + X_1 \frac{\partial \delta E_0^*}{\partial x_0} \rangle_0 \quad . \tag{V.2.10}$$

Noting Eqs. (V.2.5) and (V.2.6), Eq. (V.2.10) then reduces to

$$\langle \ddot{x}_2 \rangle_0 = -\left(\frac{q}{m\omega_0}\right)^2 \frac{\partial}{\partial c_0} |\delta E_0|^2 \quad . \tag{V.2.11}$$

Summarizing the results, we may express

$$x = x_{osc} + \delta x \quad , \tag{V.2.12}$$

where $\delta x \simeq x_1$ and x_{osc} corresponds to the long-time behaviors of the oscillating center such that

$$\ddot{x}_{osc} = -\left(\frac{q}{m\omega_0}\right)^2 \frac{\partial}{\partial x_{osc}} |\delta E_{osc}|^2 \quad . \tag{V.2.13}$$

In Eq. (V.2.13), we have, in a sense, "renormalized" Eq. (V.2.11) by evaluating its right hand side at $x_{osc} = x_0 + \langle x_2 \rangle_0$ instead of c_0. Within the framework of perturbative expansions, Eqs. (V.2.11) and (V.2.13) are, of course, equivalent to each other. Furthermore, Eq. (V.2.13) can be intepreted as the equation of motion for the oscillating center with an effective force, i.e., the pondermotive force given by

$$F_p = -\frac{q^2}{m\omega_0^2} \frac{\partial}{\partial x_{osc}} |\delta E_{osc}|^2 \quad . \tag{V.2.14}$$

In plasma physics literatures, we quite often use the terminology pondermotive potential which, obviously, is defined as

$$\phi_p = \frac{q}{m\omega_o^2} \left|\delta E_{osc}\right|^2 \quad . \tag{V.2.15}$$

Let us now examine the inequality condition, Eq. (V.2.5), more carefully. In an unmagnetized plasma, it can be interpreted either, as we did, that the particles are non-resonant or that the distance a particle travels during one wave period is much less than the corresponding wavelength. From this (adiabaticity) point of view, we can derive the oscillating-center equation of motion in the following alternative fashion. Thus, noting that

$$x = x_{osc} + \delta x \tag{V.2.12}$$

where x_{osc} is clearly defined as

$$x_{osc} = \langle x \rangle_o \quad , \tag{V.2.16}$$

and $\delta x \equiv x - x_{osc}$. Phase-averaging Eq. (V.2.1), we have

$$\ddot{x}_{osc} = \frac{q}{m}\langle \delta E + \delta E^* \rangle_0 = \frac{q}{m} \langle \delta x \frac{\partial}{\partial x_{osc}} \left(\delta E_0 + \delta E_0^*\right) \rangle_0 \quad , \tag{V.2.17}$$

and

$$\delta\ddot{x} = \frac{q}{m} \left(\delta E_0 + \delta E_0^*\right) \quad . \tag{V.2.18}$$

Here, $\delta E_0 = \delta E(x = x_{osc})$. Noting that the adiabaticity condition, Eq.

(V.2.5), ensures that x_{osc} can be held to be constant during the phase averaging, Eq. (V.2.17) can be integrated by parts to yield

$$\ddot{x}_{osc} = \langle \delta x \frac{\partial}{\partial x_{osc}} \delta \ddot{x} \rangle_0 = -\langle \delta \dot{x} \frac{\partial}{\partial x_{osc}} \delta \dot{x} \rangle_0 \quad . \tag{V.2.19}$$

Denoting $\delta V \equiv \delta \dot{x}$, Eq. (V.2.19) certainly reminds us the convective nonlinearity term, $\delta \underset{\sim}{v} \cdot \underset{\sim}{\nabla} \delta \underset{\sim}{v}$, in the usual fluid description. Furthermore, we can readily generalize this adiabatic approach to the magnetized, fully electromagnetic case. Thus, in this case, we have the equation of motion as

$$\ddot{\underset{\sim}{x}} = \frac{q}{m} \delta \underset{\sim}{E} + \frac{q}{mc} \dot{\underset{\sim}{x}} \times \underset{\sim}{B}_0 + \frac{q}{mc} \dot{\underset{\sim}{x}} \times \delta \underset{\sim}{B} \quad . \tag{V.2.20}$$

Phase averaging yields

$$\ddot{\underset{\sim}{x}}_{osc} = \frac{q}{m} \langle \delta \underset{\sim}{x}_0 \cdot \underset{\sim}{\nabla}_0 \delta \underset{\sim}{E}_0 \rangle_0 + \frac{q}{mc} \langle \delta \underset{\sim}{x}_0 \cdot \underset{\sim}{\nabla}_0 \delta \dot{\underset{\sim}{x}}_0 \rangle \times \underset{\sim}{B}_0 + \frac{q}{mc} \langle \delta \dot{\underset{\sim}{x}}_o \times \delta \underset{\sim}{B}_o \rangle_0 \quad , \tag{V.2.21}$$

and

$$\delta \ddot{\underset{\sim}{x}}_0 = \frac{q}{m} \delta \underset{\sim}{E}_0 + \frac{q}{mc} \delta \dot{\underset{\sim}{x}}_0 \times \underset{\sim}{B}_0 \quad . \tag{V.2.22}$$

In. Eq. (V.2.22), we have ignored the $\dot{\underset{\sim}{x}}_{osc} \times \delta \underset{\sim}{B}_0$ term due to the slow time variation of $\underset{\sim}{x}_{osc}$. Substituting Eq. (V.2.22) into Eq. (V.2.21), we readily find, after integration by parts,

$$\ddot{x}_{osc} = -\langle \delta \dot{\underset{\sim}{x}}_0 \cdot \underset{\sim}{\nabla}_0 \delta \dot{\underset{\sim}{x}}_0 \rangle_0 + \frac{q}{mc} \langle \delta \dot{\underset{\sim}{x}}_0 \times \delta \underset{\sim}{B}_0 \rangle_0 = -\langle \delta \underset{\sim}{v}_o \cdot \underset{\sim}{\nabla}_o \delta \underset{\sim}{v}_o \rangle_0$$

$$+ \frac{q}{mc} \langle \delta \underset{\sim}{v}_o \times \delta \underset{\sim}{B}_o \rangle_0 \quad . \tag{V.2.23}$$

Again, Eq. (V.2.23) has a transparent correspondence with the fluid description.

§V.3 Parametric Instabilities

[Ref. Advances in Plasma Physics, Vol. 6, Academic Press]

In our final topic, we shall briefly discuss the parametric instabilities; which have abundant applications in both radio-frequency wave heating and nonlinear saturation of instabilities. Specifically, parametric instabilities correspond to waves driven unstale by other waves via nonlinear mode-coupling processes. The prominent example is the nonlinear three-wave interaction. Let the three waves be denoted as $\Omega_o \equiv (\omega_o, \underset{\sim}{k}_o)$, $\Omega_1 \equiv (\omega_1, \underset{\sim}{k}_1)$, and $\Omega_s \equiv (\omega_s, \underset{\sim}{k}_s)$. In order that the nonlinear mode coupling can occur, we need to satisfy the following ω- and $\underset{\sim}{k}$-matching conditions; i.e.,

$$\omega_o = \omega_1 + \omega_s \quad , \tag{V.3.1}$$

$$\underset{\sim}{k}_o = \underset{\sim}{k}_1 + \underset{\sim}{k}_s \quad . \tag{V.3.2}$$

Here, without any loss of generality, we shall assume ω_o, ω_1, $\omega_s > 0$. Thus we take Ω_o to be the driving (pump) wave; while Ω_1 and Ω_s are the decay (daughter) waves.

Now, in the three-wave parametric decay instabilities, there are two types of decay processes. One in the resonant interaction where both Ω_1 and Ω_s are normal modes of the plasma. The other is the non-resonant interaction where either Ω_1 or Ω_s is a virtual (quasi, i.e., non-normal) mode due to, e.g., heavy damping. We shall clarify the above concept by considering the unmagnetized case; where Ω_0 and Ω_1 are the Bohm-Gross waves and Ω_s is the ion-acoustic wave. Thus, in terms of the quasi-particle jargon, we shall consider the plasmon-plasmon-phonon interaction. In this case, the resonance interaction then corresponds to a plasmon at Ω_0 decaying into a plasmon at Ω_1 and a phonon at Ω_s. Meanwhile, the matching conditions given by Eqs. (V.3.1) and (V.3.2) can be interpreted as the conservation of energy and momentum. If, however, $T_e \sim T_i$ such that the phonon is heavily ion Landau damped within the linear time scale, then only the plasmons (the Ω_0 and Ω_1 modes) exist in the nonlinear time scale. In this case, Ω_s is a virtual mode and we have a non-resonant interaction. Consequently, the decay process may be interpreted as an induced scattering of a plasmon at Ω_0 by thermal ions (i.e., $\omega_s \sim k_s v_{it}$) into another plasmon at Ω_1. Since the nonlinear wave-particle resonance condition is

$$\omega_0 - \omega_1 \simeq \left|\underset{\sim}{k}_0 - \underset{\sim}{k}_1\right| v_{it} \quad , \tag{V.3.3}$$

this decay process is also called the nonlinear ion Landau damping.

We now proceed with the detailed calculations. Let us consider only the one-dimensional case. First, we assume $|\omega_0/k_0|$, $|\omega_1/k_1| \gg v_{et}$ such that we may adopt the fluid description for the high-frequency Bohm-Gross waves. Noting that ω_0, $\omega_1 \gg \omega_s$, we can decompose the fluid quantities into its high-frequency and low-frequency components. For example, $\delta n = \delta n_h + \delta n_s$. The corresponding high-frequency equations of continuity and motion are then given by

$$\frac{\partial \delta n_h}{\partial t} + \frac{\partial}{\partial x}\left(n_0 \delta v_h + \delta n_s \delta v_h + \delta v_s\, \delta n_h\right) = 0 \quad , \tag{V.3.4}$$

and

$$\frac{\partial \delta v_h}{\partial t} + \left(\delta v_h \frac{\partial}{\partial x} \delta v_s\right) + \left(\delta v_s \frac{\partial}{\partial x} \delta v_h\right) = \frac{q}{m} \delta E_h - \frac{v_{et}^2}{n_o} \frac{\partial}{\partial x} \delta n_h \quad . \tag{V.3.5}$$

Making the perturbative expansion

$$\delta v_h = \delta v_h^{(1)} + \delta v_h^{(2)} \quad \ldots$$

and so on, and noting that

$$|\delta n_h|^{(1)} \sim \left|\frac{k_h n_o \delta v_h}{\omega_h}\right|^{(1)} \quad , \tag{V.3.6}$$

and

$$|\delta n_s|^{(1)} \sim \left|\frac{k_s n_o \delta v_s}{\omega_s}\right| \quad ; \tag{V.3.7}$$

it is then straightforward to show that

$$|\delta v_s^{(1)} \; \delta n_h^{(1)}| \sim |n_o \; \delta v_h^{(2)}| \sim O(\frac{\omega_s}{\omega_h})|\delta n_s^{(1)} \; \delta v_h^{(1)}| \quad . \tag{V.3.8}$$

That is, the dominant low-frequency modification of the high-frequency dynamics is via its density perturbation, δn_s. Equations (V.3.4) and (V.3.5) then simplify to

$$\frac{\partial}{\partial t} \delta n_h + \frac{\partial}{\partial x} \left(n_o \; \delta v_h^{(1)} + \delta n_s^{(1)} \; \delta v_h^{(1)}\right) = 0 \quad , \tag{V.3.9}$$

and

$$\frac{\partial}{\partial t}\,\delta v_h^{(1)} = \frac{q}{m}\,\delta E_h^{(1)} - \frac{v_{et}^2}{n_o}\frac{\partial}{\partial x}\,\delta n_h^{(1)} \quad . \tag{V.3.10}$$

As to the low-frequency dynamics, we need to adopt a kinetic Vlasov description in order to include the induced scattering process. The concept of ponderomotive force, therefore, becomes rather useful here. The low-frequency electron Vlasov equation is then

$$\left(\frac{\partial}{\partial t} + u\,\frac{\partial}{\partial x}\right)\delta f_{es} + \left(\frac{q}{m}\right)_e\left(\delta E_s - \frac{\partial}{\partial x}\,\delta\phi_{pe}\right)\frac{\partial}{\partial u}\,f_{oe} = 0 \quad , \tag{V.3.11}$$

with ϕ_{pe} being the ponderomotive potential for the electrons

$$\delta\phi_{pe} = -\,\frac{e}{m\omega_o^2}\,\langle\delta E_h^2\rangle_s \quad . \tag{V.3.12}$$

On the other hand, since $m_e/m_i << 1$, ion nonlinearities are small and negligible. Substituting δf_{es} and δf_{is} into the Poisson's equation, we obtain

$$\varepsilon_s\delta\phi_s = (1 + \chi_{es} + \chi_{is})\delta\phi_s = -\chi_{es}\delta\phi_{pe} \quad . \tag{V.3.13}$$

Here, χ_{js} with $j = e,i$ is the respective low-frequency linear susceptibility. From Eq. (V.3.13), we can readily derive δn_{es} as

$$4\pi e\ \delta n_{es} = k_s^2\chi_{es}(\delta\phi_s + \delta\phi_{pe}) = k_s^2\ \chi_{es}\ \frac{1+\chi_{is}}{\varepsilon_s}\ \delta\phi_{pe} \quad . \tag{V.3.14}$$

Equations (V.3.9), (V.3.10), (V.3.14), (V.3.12) plus the high-frequency Poisson's equation, thus, provide a complete set of equations; from which we can derive a modified "nonlinear" dispersion relation and, hence, parametric growth rates.

To achieve these purposes, we let

$$\delta E_h = \delta E_1 \exp(-i\omega_1 t + ik_1 x) + \delta E_o \exp(-i\omega_o t + ik_o x) + \text{c.c.} \qquad (V.3.15)$$

The $-\Omega_1$ components of Eqs. (V.3.9) and (V.3.10) are then

$$\omega_1 \delta n_1^* - k_1 n_o \delta v_1^* - k_1 \, \delta n_{se} \, \delta v_o^* = 0 \quad , \qquad (V.3.16)$$

$$\omega_1 \delta v_1^* = i \frac{e}{m} \delta E_1^* + \frac{v_{et}^2}{n_o} k_1 \, \delta n_1^* \quad . \qquad (V.3.17)$$

Using the high-frequency Poisson's equation, Eqs. (V.3.16) and (V.3.17) further reduce to

$$ik_1 \omega_1 \varepsilon_h(-\Omega_1) \delta E_1^* = 4\pi e \, k_1 \delta n_{se} \, \delta v_o^* \quad , \qquad (V.3.18)$$

with

$$\varepsilon_h(\Omega_1) = 1 - \frac{\omega_{pe}^2}{\omega_1^2} + \frac{k_1^2 v_{et}^2}{\omega_1^2} \quad . \qquad (V.3.19)$$

Meanwhile, noting that

$$\delta v_o^* \cong i \frac{e}{m\omega_o} \delta E_o^* \quad , \qquad (V.3.20)$$

and

$$\delta\phi_{pe} = -\frac{e}{m\omega_o^2} \langle \delta E_h^2 \rangle_s = -\frac{e}{m\omega_o^2} \delta E_o \, \delta E_1^* \quad , \qquad (V.3.21)$$

we have Eqs. (V.3.18) and (V.3.14) as

$$\varepsilon_h(-\Omega_1)\delta E_1^* = 4\pi e\ \delta n_{se} \frac{e\delta E_o^*}{m\omega_o^2} , \tag{V.3.22}$$

$$4\pi e\delta n_{se} = - k_s^2\ \chi_{es} \frac{(1+\chi_{is})}{\varepsilon_s} \left(\frac{e\delta E_o}{m\omega_o^2}\right) \delta E_1^* . \tag{V.3.23}$$

Equations (V.3.22) and (V.3.23) are the desired two coupled equations; which, when combined, yield the following parametric dispersion relation

$$\varepsilon_h(-\Omega_1)\varepsilon_s = -(1 + \chi_{is})k_s^2\ \chi_{es} \left|\frac{e\delta E_o}{m\omega_o^2}\right|^2 . \tag{V.3.24}$$

Let us now examine the dispersion relation assuming the Ω_1 mode is a normal mode; i.e., $\varepsilon_{hr}(\Omega_1) = 0$. Denoting $\omega_s = \omega_{sr} + i\gamma$ and

$$- \omega_1 = \omega_s - \omega_o = \omega_{sr} - \omega_o + i\gamma \equiv - \omega_{1r} + i\gamma ,$$

Eq. (V.3.24) becomes, noting $\varepsilon_h(-\Omega_1) = \varepsilon_{hr}(\Omega_1) - i\varepsilon_{hi}(\Omega_1)$,

$$-i\left(\gamma \frac{\partial\varepsilon_{hr}}{\partial\omega_{1r}} + \varepsilon_{hi}\right)\varepsilon_s = -(1 + \chi_{is})k_s^2\ \chi_{es} \left|\frac{e\delta E_o}{m\omega_o^2}\right|^2 . \tag{V.3.25}$$

For the resonant decay, we have $\varepsilon_{sr}(\omega_{sr}) = 0$. Equation (V.3.25) then reduces to

$$\left(\gamma \frac{\partial\varepsilon_{hr}}{\partial\omega_{1r}} + \varepsilon_{hi}\right)\left(\gamma \frac{\partial\varepsilon_{sr}}{\partial\omega_{sr}} + \varepsilon_{si}\right) \cong \chi_{es}^2 k_s^2 \left|\frac{e\delta E_o}{m\omega_o^2}\right|^2 \cong \left|\frac{e\delta E_o}{m\omega_o^2\lambda_{De}}\right|^2 \frac{1}{k_s^2\lambda_{De}^2} . \tag{V.3.26}$$

Setting $\gamma = 0$ in Eq. (V.3.26), we obtain the following threshold condition

$$\left|\frac{e\delta E_o}{m\omega_o\lambda_{De}}\right|^2_{th} = \varepsilon_{hi}\ \varepsilon_{si}\ k_s^2\lambda_{De}^2 \quad . \tag{V.3.27}$$

For $|\delta E_o| >> |\delta E_o|_{th}$, we have a parametric instability with the following growth rate

$$\gamma \simeq \frac{1}{k_s\lambda_{De}} \left|\frac{e\delta E_o}{m_e\omega_o^2\lambda_{De}}\right| \left(\frac{\partial\varepsilon_{hr}}{\partial\omega_{1r}} \frac{\partial\varepsilon_{sr}}{\partial\omega_{sr}}\right)^{-1/2} \quad . \tag{V.3.28}$$

In the case of the non-resonant decay, $|\varepsilon_{si}| \sim |\varepsilon_{sr}|$, the imaginary part of Eq. (V.3.25) then yields, assuming ion Landau damping dominates,

$$\gamma \frac{\partial\varepsilon_{hr}}{\partial\omega_{1r}} + \varepsilon_{hi} = -\left|k_s \frac{e\delta E_o}{m\omega_o^2}\right|^2 \chi_{es}\ \mathrm{Re}\left(i\ \frac{(1+\chi_{is})}{\varepsilon_s}\right) \simeq \left|\frac{e\delta E_o}{m\omega_o^2\lambda_{De}}\right|^2 \frac{1}{k_s^2\lambda_{De}^2} \frac{\mathrm{Im}\chi_{is}}{|\varepsilon_s|^2} \quad . \tag{V.3.29}$$

Equation (V.3.29) with $\gamma = 0$ indicates that the corresponding threshold condition is set by the high-frequency dissipation only; contrary to the resonant-decay case as shown by Eq. (V.3.27). Furthermore, since $\gamma \propto \mathrm{Im}\ \chi_{is}$, the instability mechanism is indeed associated with the ion Landau damping of the low-frequency quasi mode and, hence, the growth rate is maximized for $\omega_{sr} = \omega_o - \omega_{1r} \simeq |k_s|v_{ti}$. Noting that $k_s = k_1 - k_o$ and $|\omega_o| \simeq |\omega_1| >> |k_o, k_1|v_{ti}$, the decay is, therefore, maximal for an induced backward scattering; i.e., when $k_ok_1 < 0$.

Final Examination

(1) Consider linear ideal MHD waves in an uniformly magnetized plasma with $\underset{\sim}{B} = B_0 \underset{\sim}{e}_z$. The β value of the plasma is taken to be small; i.e., $\beta = 8\pi P_0/B_0^2 << 1$.

(1.a) Derive the linear dispersion relation.

(1.b) Show there exist seven branches of normal modes; i.e. given $\underset{\sim}{k}$, we have seven solutions of ω. [Two compressional Alfvén waves, two shear Alfvén waves, two ion sound waves, and one entropy mode].

(1.c) For the seven branches of normal modes, give their corresponding eigenvectors in terms of the seven variables (i.e., ρ, $\delta\underset{\sim}{V}$, $\delta\underset{\sim}{B}$).

(2) Let us consider resistive ($\delta\underset{\sim}{E} + \delta\underset{\sim}{V} \times \underset{\sim}{B} = \eta\delta\underset{\sim}{J}$) shear Alfvén waves in an uniform plasma with a sheared magnetic field; i.e., $\underset{\sim}{B} = B_0 (\underset{\sim}{e}_z + x/L_s \underset{\sim}{e}_y)$ and $|x| << |L_s|$ for this purpose. We shall assume, furthermore, the plasma cold and incompressible ($\underset{\sim}{\nabla}\cdot\delta\underset{\sim}{V} = 0$) and, thus, eliminate the compressional Alfvén and ion sound waves. Take pertubations of tne form

$$\delta\underset{\sim}{V}(\underset{\sim}{x},t) = \delta\hat{\underset{\sim}{V}}(x) \exp[-i\omega t + ik_y y] \quad ,$$

i.e., $k_z = 0$, hence, $k_{\parallel} = \underset{\sim}{k}\cdot\underset{\sim}{B}/B \simeq k_y x/L_s$.

(2.a) Derive the corresponding differential equation in the x space.

(2.b) Fourier transform from the x-space to the k-space, derive the following differential equation in the k-space

$$\left(\frac{d}{dk}\,\frac{(1+k^2)}{1+i\eta(1+k^2)/\omega}\,\frac{d}{dk} + \omega^2(1 + k^2)\right)\xi(k) = 0 \quad . \tag{F.1}$$

(2.c) Use Nyquist technique to prove that the wave is stable.

(3) Consider a cold, uniformly magnetized plasma with $\underset{\sim}{B} = B_0 \underset{\sim}{e}_z$.

(3.a) Show, for parallel propagating waves (i.e. $\underset{\sim}{k} = k_z \underset{\sim}{e}_z$), the dispersion relation reduces to one right hand circularly polarized wave and one left hand circularly polarized wave.

(3.b) Derive the low-frequency limits ($|\omega|/|\omega_c| \to 0$) of both waves.

(4) Lower-hybrid resonance and mode conversion.

Consider a slab plasma model in the x-z plane. Here, $\underset{\sim}{B} = B_0 \underset{\sim}{e}_z$ and x is the inhomogeneity direction; i.e., $n_0 = n_0(x)$.

For lower-hybrid waves, we have $|\Omega_e| \gg \omega \sim \omega_{pi}$ (at plasma core) $\gg \Omega_i$;

$$\left|\frac{k_\parallel}{k_\perp}\right| \sim \left(\frac{m_e}{m_i}\right)^{1/2} \quad \text{where} \quad k_\perp = k_x (\text{since } k_y = 0) \quad ;$$

$$|k_\perp \rho_e| \ll 1, \qquad \left|\frac{\omega}{k_\parallel}\right| \gg v_e, v_i \quad .$$

(4.a) Use the cold plasma description to obtain the following wave equation.

$$(\frac{\partial}{\partial x}\,\varepsilon_{\perp c}\,\frac{\partial}{\partial x} + \frac{\partial}{\partial z}\,\varepsilon_{\parallel}\,\frac{\partial}{\partial z})\delta\phi(x,z) = 0 \quad , \tag{F.2}$$

where

$$\varepsilon_{\perp c}(x) = 1 + \frac{\omega_{pe}^2(x)}{\Omega_e^2} - \frac{\omega_{pi}^2(x)}{\omega_o^2} \quad , \tag{F.3}$$

$$\varepsilon_{\parallel}(x) = 1 - \frac{\omega_{pe}^2(x)}{\omega_o^2} \quad , \tag{F.4}$$

and ω_o is the driving frequency of the r.f. source. Equation (F.3) shows the existence of a resonance at $x = x_r$ where $\varepsilon_{\perp c}(x_r) = 0$.

(4.b) As $x \to x_r$, $|k_x| \to \infty$. Since $\rho_i^2 \gg \rho_e^2$, the first microscopic effect entering is the finite ion-Larmour-radius correction. Assuming $f_{oi} = f_{Maxwellian}$, obtain this correction to $\varepsilon_\perp$; i.e., show that

$$\varepsilon_\perp \to \varepsilon_{\perp c} - \frac{\omega_{pi}^2(x)}{\omega_o^2}\,k_x^2\rho_i^2 \quad . \tag{F.5}$$

(4.c) With $(\partial/\partial z) = ik_z$ well-defined by the r.f. launcher. Plot

(4.c.1) k_x^2 vs. x for the cold plasma case.

(4.c.2) k_x^2 vs. x with FILR correction.

(5) Parametric decay instability at the lower-hybrid frequency.

Consider the resonant deay of a pump lower-hybrid (l.h) wave to another lower-hybrid wave and a slow ion sound wave. Thus, $\Omega_o = (\omega_o, \underset{\sim}{k}_o)$ is the pump l.h. wave, $\Omega_1 = (\omega_1, \underset{\sim}{k}_1)$ is the daughter l.h. wave and $\Omega_s = (\omega_s, k_s)$ is the daughter slow ion-sound wave. All waves are taken to be electrostatic. Furthermore, we have $\omega_o = \omega_1 + \omega_s$, $\omega_o \sim \omega_1 >> \omega_s$,

$$\omega_{0,1} = \omega_{lh}\left(1 + \frac{k_\parallel^2 m_i}{k_\perp^2 m_e}\right)^{1/2}_{0,1} \quad , \tag{F.6}$$

and

$$\omega_s \simeq k_{s\parallel} c_s \quad . \tag{F.7}$$

Thus, in (F.7), we have assumed $\omega_s << \Omega_i$. For further clarification, let us take $\underset{\sim}{k}_o = (k_x, 0, k_\parallel)_o$, $\underset{\sim}{k}_1 = (k_x, k_y, k_\parallel)_1$; and $|k_\perp \rho_i|$, $|k_\perp \rho_e| << 1$.

(5.a) Calculate the ponderomotive forces due to electrostatic lower-hybrid waves on electrons and ions, respectively. Noting that $\omega_{lh} \sim \omega_{pi} >> \Omega_i$, it can then be shown that electron nonlinearities dominate over ion nonlinearities. In this calculation, since the slow time scale (ω_s^{-1}) is much longer than Ω_i^{-1}, the long-time dynamics of both electrons and ions are <u>along</u> $\underset{\sim}{B}$ only.

(5.b) Obtain the parametric dispersion relation and the corresponding growth rate.

(6) Oscillating-center transformation for parametric instability.

Similar to the guiding-center transformation used in treating magnetized plasmas, the existence of oscillating centers suggests that an analogous transformation should be useful in treating parametric (i.e., nonlinear mode-coupling) processes. We shall illustrate this point by considering unmagnetized plasmas in a dipole (i.e., spatially uniform) pump;

$$\underset{\sim}{E}_0(x,t) = \underset{\sim}{E}_0 \cos \omega_0 t \quad . \tag{F.8}$$

The Vlasov equation then is

$$\left[\frac{\partial}{\partial t} + \underset{\sim}{v} \cdot \frac{\partial}{\partial x} + \frac{q}{m}\left(\underset{\sim}{E}_0 + \delta\underset{\sim}{E} \cdot \frac{\partial}{\partial \underset{\sim}{v}}\right)\right] f = 0 \quad . \tag{F.9}$$

The transformation is then defined as, for $j = e,i$,

$$\underset{\sim}{V}_j = \underset{\sim}{v} + \underset{\sim}{u}_j \quad , \quad \underset{\sim}{X}_j = \underset{\sim}{x} + \underset{\sim}{\rho}_j \quad , \tag{F.10}$$

where

$$\dot{\underset{\sim}{u}}_j = \left(\frac{q}{m}\right)_j \underset{\sim}{E}_0 \quad \text{and} \quad \dot{\underset{\sim}{\rho}}_j = \underset{\sim}{u}_j \quad . \tag{F.11}$$

(6.a) Show that the linearized Eq. (F.9) transforms to

$$\left(\frac{\partial}{\partial t} + \underset{\sim}{V} \cdot \frac{\partial}{\partial \underset{\sim}{X}}\right)\delta\hat{f} = - \frac{q}{m}\, \delta\hat{\underset{\sim}{E}} \cdot \frac{\partial}{\partial \underset{\sim}{V}} \hat{f}_0 \quad . \tag{F.12}$$

Here, $\hat{f} = \hat{f}(\underset{\sim}{V}, \underset{\sim}{X})$ and the subscript j is suppressed here.

(6.b) Following the analogy with the guiding-center treatment, try to show that Eqs. (F.10), (F.11), and (F.12) and the Poisson's equation lead to the following coupled set of equations.

$$D_{e,n}E_n = -\chi_{e,n} \sum_{p=-\infty}^{\infty} J_{n-p} I_p \quad , \tag{F.13}$$

$$D_{i,n} I_n = -\chi_{i,n} \sum_{p=-\infty}^{\infty} J_{p-n} E_p \quad . \tag{F.14}$$

Here, n = integers, $\chi_{j,n} = \chi_j(\omega + n\omega_o, \underset{\sim}{k})$ is the jth species susceptibility. $D_{j,n} = 1 + \chi_{j,n}$, J_m is the Bessel function with the argument $\mu = \left|\underset{\sim}{k}\cdot(\underset{\sim}{\rho}_e - \underset{\sim}{\rho}_i)\right|$. [Note that in the no-pump limit $\mu \to 0$, we do recover the usual disperson relation $1 + \chi_{e,n} + \chi_{i,n} = 0$.] Equations (F.13) and (F.14) show an interesting features; i.e, there is <u>no</u> parametric process if $m_i \to \infty$ i.e., $|\chi_i| \to 0$.

(6.c) Examine Eqs. (F.13) and (F.14) in the weak-pump limit i.e.; $\mu \ll 1$. Can you recover the parametric dispersion relation for the plasmon → plasmon + phonon decay discussed in class? For this purpose, take $\omega = \omega_s$, $\omega + \omega_o = \omega_1$, $\underset{\sim}{k}_s = \underset{\sim}{k}_1$.

(6.d) A comment: this transformation technique can be straightforwardly generalized to magnetized plasmas.

(7) Quasi-linear theory.

In Kaufman's treatment of the quasi-linear theory, the phase-mixing argument is crucial because he is interested in the case that the normal modes may be <u>weakly</u> damped (i.e., $\gamma_k < 0$, $|\gamma_k| \ll |\Delta k\ v_t|$). However, if the normal modes are <u>unstable</u> such that, for $t \sim \gamma_k^{-1}$, the normal modes dominate (i.e., the initial perturbations are negligible in any case), then Kaufman's results can be derived in a much simpler fashion. Please show it.